Molecular biology and crop improvement

T0297156

Commission of the European Communities (Division for Genetics and Biotechnology of the Biology, Radiation Protection and Medical Research Directorate, Directorate-General for Science, Research and Development)

Molecular biology and crop improvement

A CASE STUDY OF WHEAT, OILSEED RAPE AND FABA BEANS

R.B. Austin
with
R.B. Flavell, I.E. Henson and H.J.B. Lowe
Plant Breeding Institute, Trumpington, Cambridge

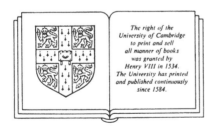

The right of the
University of Cambridge
to print and sell
all manner of books
was granted by
Henry VIII in 1534.
The University has printed
and published continuously
since 1584.

CAMBRIDGE UNIVERSITY PRESS

Cambridge

London   New York   New Rochelle

Melbourne   Sydney

CAMBRIDGE UNIVERSITY PRESS
Cambridge, New York, Melbourne, Madrid, Cape Town, Singapore, São Paulo, Delhi

Cambridge University Press
The Edinburgh Building, Cambridge CB2 8RU, UK

Published in the United States of America by Cambridge University Press, New York

www.cambridge.org
Information on this title: www.cambridge.org/9780521112918

First published 1986
This digitally printed version 2009

A catalogue record for this publication is available from the British Library

ISBN 978-0-521-32725-1 hardback
ISBN 978-0-521-11291-8 paperback

CONTENTS

PREFACE

This book is the report of a study on opportunities for the application of molecular biology to crop improvement in the EEC, with particular reference to wheat, oilseed rape and faba bean, carried out by the Plant Breeding Institute, Cambridge, United Kingdom, under contract to the Commission of the European Communities (Division for Genetics and Biotechnology).

The Division called for an assessment of the opportunities for the application of molecular biology to the improvement of three species which play an important role in European agriculture, namely wheat, oilseed rape and faba beans. We considered that an assessment of this kind should be based on existing knowledge of these crops. From this, the need for particular improvements could be identified and an assessment made of the opportunity for making them offered by the available techniques of molecular and cell biology.

The main conclusions from the study are:

The biochemical study of genes and their primary products will continue to be of great value for research in plant biology. Through such research, a much better understanding will be gained of the molecular basis of growth and differentiation, of responses to environmental factors and of host–pathogen and host–pest interactions.

The ability to insert foreign and modified genes into plants, when better developed, will offer entirely new opportunities for making defined, limited changes to plant genotypes. Comparison of the modified and original genotypes will enable unequivocal tests to be made of alleged limiting points in plant metabolism and of the basis of resistance to particular pests and pathogens. This work will also show how improvements in yield, quality and pest and pathogen resistance may be achieved.

The techniques of molecular biology will not replace existing plant breeding practice. Comprehensive testing and evaluation for yield, quality and pest and disease resistance will be needed in order to evaluate any products of genetic engineering, just as it is needed for evaluating the progenies of crosses made as part of conventional breeding programmes.

For wheat there is an urgent need to find and develop a vector for transferring alien genes to the species. Subjects for study by molecular biological methods include: photosynthesis, drought resistance and the basis of resistance to the major pathogens, namely yellow and brown rust, mildew and take-all.

For oilseed rape, rapid means of identifying S-incompatibility alleles at the seedling stage, probably by cDNA probes, would aid the development of F_1 hybrid varieties. The possibility of transferring genes from the related genus *Moricandia* to reduce photorespiratory losses in oilseed rape should be investigated. The molecular basis of resistance to stem canker should be studied. Generalised resistance to the numerous insect pests of oilseed rape (many of which are common to other brassica crops) should be sought and transferred to the crop.

For faba beans, where the normal process of sexual hybridisation cannot be used to transfer genes from related species, alternative methods of gene transfer are required. Attention should be given to the transfer of genes which would increase drought resistance and of those which would confer improved resistance to the fungal pathogens *Ascochyta* and *Botrytis* and to the aphid *Aphis fabae*. The scope for modifying the storage proteins and increasing their amounts in the seeds needs to be evaluated. Increased autofertility would reduce or avoid fluctuations in yield caused by the variable effectiveness of pollination by bees. Improved, more stable cytoplasmic male sterility would make the production of F_1 hybrid faba beans a practical possibility.

We have not attempted to provide a comprehensive reference list. The biology of the three crops, together with information on crop quality and pests and diseases, and the basic principles of plant breeding and of molecular biology are covered in many textbooks. Recent developments in the biology of the three crops are covered in reviews and conference proceedings, each of which contains a comprehensive reference list. A reading list which gives this information is given at the end of the book.

Statistics on production and consumption are generally for the ten European Community countries and have been taken from the Eurostat quarterly summary of crop production statistics, published by the Statistical Office of the European Communities, Luxembourg.

December 1985   R.B. Austin
                   R.B. Flavell, I.E. Henson, H.J.B. Lowe

ACKNOWLEDGEMENTS

We are indebted to numerous colleagues for their help with this book. Particular thanks are due to the plant breeders at the Plant Breeding Institute, J. Bingham, FRS, K.F. Thompson and D.A. Bond; to P.R. Day and P.I. Payne for their helpful suggestions on the applications of molecular biology; to R. Johnson, P.R. Scott and M.S. Wolfe for their guidance on the sections on diseases; and to A.R. Stone of Rothamsted Experimental Station for guidance on the sections on plant nematodes. Valuable comments were made on an early draft of the book by Dr B.J. Miflin of Rothamsted Experimental Station, Professor F. Quétier of the Université de Paris-Sud, Professor A. Rörsch of the Rijksunversiteit te Leiden and Professor D. von Wettstein of the Carlsberg Laboratory. The support and encouragement of the staff of the Division for Genetics and Biotechnology of the Commission of the European Communities is also gratefully acknowledged.

The care and patience of Mrs Karen Parr and Miss Sue Land, who typed the drafts of the study is gratefully acknowledged.

1
Introduction

During the last two decades there has been a dramatic increase in our understanding of the chemistry and biology of nucleic acids. It is now possible to isolate individual genes and determine their structure, to modify the structure in a directed way and, for some organisms, to achieve a particular change in the function of the organism, with the gene active either in its 'natural' background or in a different organism.

It is widely believed that these techniques can be applied to crop plants, greatly to aid the production of improved varieties. It is the aim of this report to examine in detail for wheat, oilseed rape and faba beans the objectives of breeders of these crops in the EEC and to assess the scope for applying the techniques of molecular biology to achieve these objectives.

Taking a long perspective, it is known that for several millenia man has selected crops for ease of cultivation and harvest, greater yield, appropriate quality and resistance to pests and diseases. Scientific breeding, involving deliberate hybridisation followed by selection of the segregating progeny for particular traits of interest, has, with few exceptions, been carried out for less than a century.

Over this period there have been major changes in agricultural practices, notably the much greater use of fertilisers and the introduction of selective herbicides and improved pesticides and fungicides. These changes have relieved some constraints on yield and breeders have responded by modifying the form and growth patterns of their crops so that they can better exploit the improved environment. For winter wheat in the UK, for example, yields have increased from about 2.5 t ha^{-1} during the first decades of this century to their present level of 6–7 t ha^{-1}.

Taking an evolutionary view, it may be supposed that up to the beginning of the twentieth century, crop varieties became fitted to exploit optimally the prevailing agricultural environment. It may also be supposed that, with the exception of nitrogenous fertilisers, the agricultural environment has now reached a new, near ideal level. In due course, depending on the effort devoted to breeding, varieties will be produced which optimally exploit this new environment, so that a new yield plateau will be reached. If this concept is correct, the breeders' task will then be to maintain or improve resistance to diseases and pests, thus decreasing the costs of production, and to respond to any change demanded by millers and bakers or other users of wheat by genetically varying the composition of the grain. A major uncertainty in the above argument is whether, for environmental protection and other reasons, the use of nitrogenous fertilisers will reach a plateau and what this level will be.

The extent to which increased yields are perceived to be required in the countries of the EEC will determine the effort devoted to breeding new varieties. It is not the purpose of this report to make a detailed analysis of this problem. However, any plan for crop improvement, if it is to be credible, must take into account the need for increased yield and modified quality, as perceived from the study of past trends and projected needs. In the EEC, agricultural policy is determined by political and social, as much as by economic and biological considerations. However, in the long term, costs of production must be competitive with those of non-EEC producers. Present policy has the effect of encouraging the production of indigenous, well-adapted crops and discouraging their importation from non-EEC countries. The advantages of the policy are social – slowing down rural de-population, and strategic – avoiding dependence on non-EEC sources. However, as currently applied, the policy results in over-production of some commodities. The surpluses have to be exported to non-member countries at a cost to the EEC. As yields vary from year to year, to guarantee self-sufficiency, some over-production is inevitable. In the long term, it may be expected that unless costs of production can be reduced so that it is economic to grow for export, agricultural support will be limited to ensuring that production can be guaranteed to meet EEC demand.

For one commodity, protein concentrate for animal feeding stuffs, the EEC is a substantial importer. This shortfall could be met by increasing the production of grain legumes, including peas, faba beans, soya beans and lupins, though industrial methods for producing protein from hydrocarbon or carbohydrate feedstocks and nitrate or ammonia could partly meet the demand.

In the longer term, recognition of the need to produce more timber in the EEC may encourage the planting of more forests, decreasing the

2

area available for arable crops. If the area under arable crops does decline, higher yields will be required to maintain production.

The interpretation of EEC agricultural policy and future trends outlined above has the following implications for crop improvement policies. First, to become more competitive with major world producers, net costs of production need to be reduced. Crop improvement can help to achieve this by providing varieties which are easier to grow and are less dependent on agrochemicals. However, as the fixed costs of production are much greater than the variable costs (which include agrochemicals and fertilisers), an increased yield per hectare is still, at least on moderately fertile land, a major means of decreasing the cost per unit of production. Second, there is scope for expanding the production of crops, particularly grain legumes and sunflower, where presently there is significant import, which would substitute for imports. In addition to these policies, it would generally be accepted as biologically and economically desirable to maintain a reasonably diverse agriculture, in terms of the species cultivated, to lessen the economic and social effects of a serious epidemic of a pest or disease, or of adverse weather. For illustration, it may be prudent to maintain the breeding of varieties of oats and rye, as well as a modest area devoted to these crops, so that in the event of a catastrophic reduction in wheat or barley production due to disease, the production of rye and oats could be increased to meet the shortfall.

Finally, new uses for existing crops may be discovered which could stimulate production. Crop composition could be modified so that the crops could be grown to produce high-value biochemicals. Crops could also be used as feedstocks for chemical and biochemical processes.

Chapters 4, 5 and 6 of this report analyse the opportunities for the application of molecular biology to the improvement of wheat, oilseed rape and faba beans. The analyses take account of what are perceived to be the needs of EEC agriculture. First, the need for modifying particular features of the crops is considered. Second, ways in which each feature might be modified are discussed. Where a method based on molecular biology is deemed to be the most appropriate of various methods, or is the only method, this view is noted.

The EEC has already funded a programme of research on 'Biomolecular Engineering'. A major objective of that part of the programme concerned with crops was to develop methods for the transfer of genes, thus augmenting the large effort devoted to this end in laboratories throughout the world. To provide a framework for assessing the feasibility of achieving the objectives described in Chapters 4–6, Chapter 2 considers the technology currently used by plant breeders, and some of its limitations, and Chapter 3 summarises the techniques of molecular biology which are currently available and could be applied to improve these crops.

2

Crop improvement by breeding

2.1   Traditional methods

The technology of plant breeding has developed from the science of genetics. But as crop improvement by breeding depends on recognition of particular traits needed for achieving high and stable yield, pest and disease resistance and quality, success in breeding also depends on an understanding of plant physiology, pathology and biochemistry.

The fundamental concept of genetics is the gene, the unit of inheritance. Each gene controls or influences some aspect of plant behaviour and the gene complement, or genome of a plant consists of 10^4–10^5 genes. A proportion of the genes can exist in more than one form, or allele, at any given locus, and so individuals of a species with the same loci, but with different allelic variants, will be different in their form or function. Many genes are common to most organisms. For a given species, an even greater proportion of the genes will be the same for all individuals, and only a relatively small proportion will display allelic variation. Traditional plant breeding is concerned mainly with the directed reassortment of the allelic variants to produce a combination, or genotype, which best approaches a supposed ideal combination. The allelic variants of many genes have very small effects (i.e. the difference between the alleles of a particular gene have only minor consequences for plant form or function), but some have larger easily recognisable effects. Loci which have allelic forms with very different effects are relatively easy to manipulate by crossing and selection. However, in breeding programmes, as the number (n) of genes with major effects for which selection needs to be practised increases, the number of plants of a segregating generation which has to be grown in order to be able to find the one with the desired allelic composition increases by a function of 2^n depending on the inheritance of the character and the generation in which selection is practised. The genetics of plant characters for which breeders need to select is often complex and not well understood, but the same general argument applies – as the number of characters for which explicit selection needs to be practised increases, so the number of individuals in a segregating generation which needs to be examined to have a reasonable probability of detecting the one of interest increases as a function of 2^n. This simple model applies for characters where the variation of interest is caused by genes with major effects, where there are two alleles per gene, and where there are no interactions between genes. In reality there are multiple alleles at some loci, dominance of some alleles and interactions between genes, and the expression of genes depends on the environment. These are additional reasons why breeders need to work with large populations and why they have at present little alternative to the traditional method of selecting for what they believe is the ideal plant type (or ideotype) for maximum yield.

These considerations also apply to pest and disease resistance and the components of quality, though generally to a lesser degree because fewer genes are involved. However, the recognition of pest and disease resistance is greatly facilitated by screening tests. Similarly, where quality is an easily measured trait, efficient screening tests can be devised.

The task of identifying desirable genotypes in segregating populations presupposes that variation for the characters of interest is present in the populations. If it is not, the transfer of desired genes to the crop species from a related species is required.

Analysis of progress in plant breeding shows that, given a sufficient scale of operation, new varieties can be produced which are better adapted to the prevailing agricultural environment. With high-input agriculture, this has meant much improved yield, especially notable in Western Europe where water is not usually a dominant limitation. Simple economic analysis also shows that plant breeding programmes for the major crops are very cost effective. Furthermore, for many crops, breeders believe that there is still much variation which remains to be exploited, so they are optimistic that progress can be maintained for some decades. It will be aided by maintaining the scale of the programmes and increasing their efficiency. Some of the improvements in efficiency, such as decreasing the time from generation to generation, improving the accuracy of field trials and identifying disease resistance more rapidly, do not depend on new technology but on the refinement of existing technology. It is anticipated that molecular biology will enable the efficiency of breeding programmes to be increased further still, and that it will also make it possible to achieve hitherto unattainable objectives.

2.2   Improvements in breeding technology

Improvements in breeding technology can be considered under several categories:

1. The introduction of alien variation
2. An increase in the precision of selection
3. An increase in the speed of selection
4. Modification of the breeding system
5. A decrease in the generation time
6. An improved and more precise definition of breeding objectives

2.2.1   Introduction of alien variation

The only well-established method for introducing alien variation is sexual hybridisation. Various procedures are employed to overcome the natural barriers to hybridisation, but there are some species, including *Vicia faba*, where it has not so far been possible to produce hybrids with related species. Although repeated backcrossing and selection can result in the transfer of only the desired character, the method is successful only where an easily identifiable gene is involved. The transfer of a limited number of genes between related species may be possible with irradiated pollen, but such transfer is random. Molecular biology could aid the introduction of alien variation by making possible the isolation of particular genes from an alien source, not necessarily a related species, and their incorporation into the species of interest, either by a vector or possibly by direct injection. Where it is desired to explore the effects of transfer of a larger proportion of the genome, the use of irradiated pollen and somatic hybridisation achieved by fusion of isolated protoplasts may be useful procedures.

2.2.2   Increase in the precision and speed of selection

At present, selection is based on examination of entire individuals, or in some cases organs excised from them, and identification of the character of interest visually or chemically. Where it is possible to recognise a gene of interest or its product, preferably non-destructively, selection can be carried out with 100 per cent efficiency.

For example, analysis of part of the endosperm of a wheat grain can reveal the presence or absence of particular subunits of the glutenin protein. Individuals carrying desired combinations of subunits can be identified at an early stage in the breeding programme. In the future, it may be feasible to identify particular alleles with cDNA probes, or the products of the alleles with probes based on antibodies to the products. It may be envisaged that such tests could be carried out on large numbers of seedlings and only those showing a positive reaction retained. Such procedures would also increase the speed of selection, and could greatly reduce costs as only the selected individuals would need to be grown to maturity.

2.2.3 Modification of the breeding system

The breeding system of a crop species (i.e. whether it is naturally self-pollinated or cross-pollinated) determines the strategy a breeder has to use to produce new varieties. Other features being similar, self-pollinated species are easier to breed than cross-pollinated ones. There is a case, therefore, for converting cross-pollinated species to the self-pollinated habit. Achievement of this involves more than just selection for autofertility, because the genetic architecture of a cross-pollinated species can make it unfit for the degree of homozygosity imposed by the self-pollinated, autofertile habit. Provided that they could be identified, molecular biology could help in the transfer of genes for self-fertility and in the identification and elimination of any major genes associated with the cross-pollinated condition. The molecular basis of self-incompatibility is not adequately understood and much more research may be needed before breeding systems can be modified at will. For example, self-incompatibility may result from the presence of genetic information for self-recognition. If so, mutation breeding may be more effective than gene transfer as a means of converting self-incompatible species to self-compatible, inbreeding, forms. On the other hand, some work suggests that several unlinked loci are required for the expression of self-incompatibility alleles, and where this is true, the transfer of several genes may be required. The identification and assembly of these genes into a foreign host could be achieved only by molecular biological methods.

2.2.4 Decrease in the generation time

Given unlimited resources of land and manpower, progress in breeding is limited by the time which elapses between one generation and the next. This is most serious for perennials, notably trees, where, in addition, many of the characters of interest cannot be assessed early in the life cycle. For many crops, including wheat, oilseed rape and some trees, breeders are able to reduce considerably the duration of the life cycle (time from seed to seed) by growing plants in suitable environments. By this method three to four generations of spring wheat and two to three generations of winter wheat can be raised each year, compared to the single generation per year for the same plants when grown in the field as crops. For apples, the average generation time is

8

up to eight years but it can be reduced to two to three years by manipulation of the environment. Improved knowledge of the molecular basis of flower induction may enable the process of flower initiation to be 'switched-on' by treatment of seedlings with chemicals so reducing even further the generation time.

2.2.5 *Improved and more precise definition of breeding objectives*

As already noted, breeders' objectives are generally formulated in terms of component characters which together constitute an ideotype. Lack of knowledge, or logistical limitations, generally prevent the ideotype being defined in terms of particular genes. For some qualitative, 'presence or absence', characters, the ideotype can be defined precisely. As, through the application of molecular biology, more is learnt of the genetic control of processes important for growth and yield, the number of genes which can be recognised to have important qualitative or quantitative effects will grow. It will be possible to define ideotypes with greater precision and breeders will have the means for selecting for the components of the ideotype at an early stage in the breeding process.

The cost of breeding programmes limits the range of types of variety that it is economic to produce, even though the need for different types is recognised. For example, different types of wheat are needed for use in animal feeding-stuffs, for bread-making and for biscuit making. For the best yield, different types are needed for different climatic and edaphic areas of the EEC. If, for example, eight such zones were recognised, up to $3 \times 8 = 24$ types of varieties would be needed. If the number of genes involved in determining adaptation and quality was sufficiently small, their transfer by molecular biological methods, as distinct from back-crossing, could make the production of a range of variety types a feasible proposition, and be of considerable service to agriculture.

In summary, for each of the six aspects of breeding technology considered above the challenge to molecular biologists is to devise effective and logistically feasible procedures which will enable breeders to respond quickly and effectively to the constantly changing needs of agriculture. The adoption of new techniques in commerce is of course partly dependent on general economic circumstances.

2.3 The physiological basis of yield

Yield is the integration of very many component processes, each influenced by genotype and environment. Most of these processes are common to all plants and the genes that code for the molecular structures which carry out these processes are highly conserved. It follows that variation at the genus level is likely to be determined by allelic variation

9

at a relatively few loci (tens or hundreds, rather than thousands), as well as by changes in the location of genes brought about by more gross changes in chromosomal organisation, particularly inversion and deletion. On the other hand, especially at the ecotype and cultivar level, variation is most likely to be due to allelic differences at loci coding for regulatory molecules affecting development and metabolism. At present the genes at these loci are much more difficult to identify than those which code for molecules produced in much greater abundance.

A necessary feature of plants is their ability to compensate for the effects of environmental and genetic changes. Adverse environmental conditions can be compensated by subsequent changes in the rates or durations of processes other than those originally affected. For example, low light intensity reduces photosynthetic rate but dry matter yields are proportionately less affected because low light intensity results in greater leaf expansion and larger photosynthetic unit size, both of which are changes that increase the quantum yield and so compensate in part for the reduced light. Genetic changes can be compensated, or buffered, by metabolic 'interlocking'. Thus a genetic change in a particular step in a complex metabolic pathway may alter the concentrations of metabolites of the pathway but feedback effects may occur on the rates of related reactions in the pathway, with the net result that there are small or negligible changes in the operation of the entire pathway. It can thus be seen why it is so difficult to identify and manipulate 'yield' genes, and why breeding for high yield has relied so much on 'accelerated natural selection'.

From our present knowledge of physiology and biochemistry it is possible to construct blueprints, or ideotypes, for the optimal performance of a crop in particular conditions. These ideotypes are couched in terms of crop structure and of processes at the organ level. However, as argued above and elsewhere in this report, the complexity of plants and their interactions with the environment is such that no ideotype can at present be regarded as definitive. The effects of particular features of crop structure or of changes in processes need to be determined by experiment for each crop species in the relevant range of environments. Such experiments, which involve the preparation of defined genotypes and their testing in carefully controlled field experiments, are expensive in time and manpower. However, the process of defining an ideotype, testing its elements, modifying the ideotype, and so on, is a logical approach to the application of physiology to crop improvement by breeding. At the other extreme, the breeder would pay no attention to crop structure or physiology and select only for yield. The latter procedure, whilst avoiding errors based on prejudice about the significance of component characters, is likely to be rewarding only in the first decades of breeding programmes, and of diminishing value subsequently.

In practice, progress in breeding the major crops appears to have increased with time, rather than decreased. This can be attributed partly to the increased scale of breeding programmes, partly to new opportunities created by improvements in the agricultural environment (e.g. herbicides, fertilisers) and partly to improvements in the efficiency of the breeding processes. Ultimately, progress will depend on improved knowledge of plant form and function, from which improved ideotypes can be defined and better selection techniques devised.

2.4   Resistance to disease

Plant pathogens have the potential to cause major yield losses in most agricultural crops. Presently, control is achieved partly by breeding resistant varieties and partly by applying chemicals to the crops. Chemical control can be expensive, hazardous to the operator and damaging to the environment, and is not always effective.

Modern agricultural methods may intensify disease problems, owing to:

1. cultivation of large areas with a single crop
2. reliance on a few cultivars or a single cultivar
3. reduction in natural predator/competitor populations due to use of chemicals or destruction of habitats
4. continuous cultivation of the same crop over many years

The importance of using varieties possessing a suitable level of disease resistance is widely appreciated and evident both in the level of resistance which is required before an otherwise suitable variety of a crop is accepted into agriculture, and by the demise of established varieties when resistance is diminished or lost due to changes in pathogen populations. It follows that the incorporation of a suitable level of resistance to major diseases is a major preoccupation of plant breeders, involving much time and effort in screening and evaluation for disease resistance.

Breeding for resistance is also a continuing need. For many diseases the pathogens exist as physiologically distinct races displaying cultivar specificity. General resistance to all races is often difficult to achieve and, due to short generation times and high rates of increase, populations of pathogen races to which a new cultivar is susceptible can evolve rapidly.

There is an urgent need for a better understanding of the nature of host – pathogen interactions so that new methods can be devised for achieving improved and more durable resistance of crop plants to diseases. Molecular biology will play an essential role in enabling present methods to be improved, as described below.

2.4.1 Theories of disease resistance

In studies on the nature of disease resistance in plants, most attention has been given to variations in the host and pathogen which determine the species and cultivar specificity of a given pathogen or particular races of it. In the following discussion, which deals mainly with these aspects, it should be borne in mind that most plants are resistant to most potential pathogens. Because the resistance is usually absolute, it is very difficult to determine its basis. It is tempting to suppose that if the genes involved could be transferred to crop plants, one or a few genes could provide permanent resistance to a given pathogen of a crop, providing there were no adverse side-effects. As a long-term goal, this appears to offer a greater reward than seeking to increase resistance by exploiting 'micro-variation' among the varieties of a crop, or even among related species.

It is sometimes claimed that high yield inevitably carries with it the penalty of susceptibility to disease (i.e. that the metabolic costs of resistance, if eliminated, would permit higher yield). However, there is little direct evidence on this question and it is more widely accepted that the growing of crops as monocultures promotes the multiplication and evolution of potential pathogens, and that this, rather than the metabolic cost is the more probable reason why diseases are so troublesome in crops. In some cases, especially where insect (see Sections 3.8 and 5.3.5) and other animal pests are concerned, the greater susceptibility of crop plants may be associated with the elimination of constituents that are toxic to humans.

Considerable information exists, especially for wheat, on the number and nature of genes conferring resistance, particularly to fungal biotrophs such as rusts and powdery mildew. From such information, together with rather limited biochemical evidence, general mechanisms of disease resistance have been postulated. Somewhat different mechanisms are necessary to account for the species specificity of pathogens as opposed to the cultivar specificity which is usually the major concern of the plant breeder.

Cultivar specificity. Any model explaining cultivar specificity has to take account of the following generalisations:
   1. genes conferring race-specific pathogen resistance in plants are usually dominant and resistance is often simply inherited

12

2. pathogen avirulence genes are usually dominant and simply inherited
3. a matching of dominant resistance genes and avirulence genes is required for incompatibility; all other combinations result in compatibility (i.e. susceptibility of the host)
4. cross-protection is sometimes possible against a compatible (virulent) pathogen by prior infection with an incompatible (avirulent) pathogen
5. expression of a resistance gene is affected by the genetic background of the host – genes which suppress the action of particular resistance genes occur
6. some pathogen virulence genes appear to confer a selective disadvantage when present unnecessarily (i.e. when the host lacks the corresponding resistance gene) in the population, this serving to restrict their frequency (the corollary is that avirulence gene products may have other functions than the conferring of avirulence)
7. some resistance genes have multiple alleles
8. some resistance genes are temperature-dependent in action
9. close contact between pathogen and host appears to be necessary for induction of defence reactions

In cultivar specificity, host defences against an avirulent race (a race to which the cultivar is resistant) are thus assumed to be positively activated as a result of the interaction between a plant receptor molecule (assumed to be a product of the resistance gene), and a race-specific pathogen elicitor molecule.

A combination of plant and pathogen in which one or both of these molecules are absent results in a failure to induce host plant defences and hence in susceptibility. This situation forms the basis of the gene-for-gene relationship between pathogen and host. In some cases resistance may be an active process and is induced following infection. In others, some avirulence and resistance gene products may be constitutive, albeit tissue or cell-specific. Clearly, constitutive defences associated with morphological features, surface topography, cuticle composition and structure, and other anatomical barriers to infection are seen to be additional to, and not part of, those defensive reactions induced in gene-for-gene interactions. This implies that 'resistance' genes are of at least two types: those involved in specific recognition of invading pathogens and those controlling less specific, largely constitutive features. The latter may be more important during the initial stages of infection (spore adhesion, germination, germ tube growth, appressorium formation and penetration), while specific gene-for-gene resistance genes may be more important at later stages.

Following an initial recognition event (the binding of an elicitor to a receptor at the pathogen/host interface), a 'second messenger' is postulated to initiate responses in the host nucleus and perhaps at other sites leading to activation of a defence mechanism or mechanisms. These may be of several kinds, involving phytoalexin production, lignification of cell walls and associated processes resulting in the 'hypersensitive reaction' (HR) in which cell necrosis occurs , usually in a localised zone surrounding the infection site. The appearance of such necrotic lesions is often characteristic of particular resistance gene/pathogen interactions. The importance of cell necrosis, or HR, in inhibiting pathogen development is not always clear, however, as inhibition of fungal growth can occur either before or after HR depending on the pathogen–plant system.

Phytoalexin production has received much attention. Phytoalexins are low molecular weight products of phenyl-propanoid metabolism that are antifungal compounds induced following infection. However, they can be elicited non-specifically by chemical or physical factors in the absence of disease organisms. The involvement of phytoalexins has yet to be demonstrated for several important cereal–pathogen interactions.

As yet, neither elicitor nor receptor molecules have been isolated and characterised although several attempts have been made and there is much circumstantial evidence concerning the likely chemical nature of the molecules. The identity of the postulated secondary messenger appears entirely unknown, although chemicals serving analogous functions in higher animals have been well researched. The molecular events initiated by the messenger are likewise little understood. Inhibitors of RNA and protein synthesis interfere with the development of resistance reactions. Therefore, de novo synthesis of specific mRNAs and proteins involved, for example, in phytoalexin synthesis, probably occurs in response to infection.

Species specificity. Clearly, it is difficult to account for species specificity in terms of an interaction of elicitor and receptor similar to that postulated for cultivar specificity. It is unlikely that a plant could produce receptors specific for all potential invaders. Rather, species-level compatibility is achieved as a result of positive attributes of the pathogen which act to overcome or perhaps suppress innate host defences. Physical barriers to infection, mentioned earlier, must play a role in this. However, the fact that plant defence responses can be overcome or suppressed in one case (species level) while being induced in another (cultivar level),

14

appears a paradox. It may be explained by the observation that pathogens are usually narrowly specialised with respect to host species and cultivar specificity is superimposed on this. Attempts to reconcile this have been made. Although breeders have been understandably concerned with cultivar specificity, acquisition of general resistance implicit in species-level interactions should, as noted above, perhaps be an ultimate goal.

Although certain crucial evidence is lacking, current concepts of the mode of action of resistance genes serve as a useful basis for future research. Many questions remain unanswered. These include: unifying species-specific and cultivar-specific mechanisms, reconciling race-specific induction and non-specific elicitation of, for example, phytoalexin production, and accounting for the action of host-specific pathogen toxins which mimic disease symptoms.

2.4.2 Present techniques for obtaining disease resistance and their limitations

Current methods of breeding for disease resistance are largely empirical. However, an appreciable knowledge of the genetics of resistance exists, particularly for the aerially disseminated biotrophic diseases of cereals (rusts, mildews). This knowledge enables rational choices to be made concerning the suitability of parents and the use of sources of resistance in related species.

Selection for resistance generally depends on exposure of the plant material to appropriate pathogen populations and levels, usually under field conditions. To select for resistance relevant to current agriculture the pathogen populations used must reflect those likely to be met with under farming conditions. On the other hand, to define the nature of the resistance precisely and to monitor the inheritance of specific resistance genes, well characterised, 'pure' races of pathogens need to be used.

Environmental conditions can greatly influence the outcome of specific host–pathogen interactions, the progress of infection and the expression of certain resistance genes. The age and stage of development of the host plant often determine the expression of resistance, thereby complicating the screening procedure.

2.4.3 Other ways of decreasing the incidence of disease

In addition to increasing the resistance of crops to diseases there is scope for the alteration of agronomic practices and cropping patterns so that the opportunities for infection are reduced and the trends described above are counteracted. To this must be added the development of improved fungicides. Whilst this subject is beyond the scope of this report, it should be noted that an improved understanding of particular host–pathogen interactions may help in the design of more effective fungicides than are at present available. Some may be the products of plant genes and be best made by microorganisms with the appropriate higher plant genes inserted into them.

2.5   Resistance to insect pests

Over the last forty years, insect pests have been controlled mainly by the application of insecticides. Much research effort has been devoted to the development of new insecticides and of methods for using them. In contrast, the control of pests by exploiting genetic resistance has received less attention.

Defence against pests and pathogens is a necessary property of plants, allowing them to survive in nature. Physical features, for example epidermal hairs and dense, fibrous tissues, as well as chemicals and also the patterns and timing of plant growth can all contribute to resistance against pests.

A defensive role can be ascribed to many secondary metabolites, most of which are formed by biosynthetic pathways involving several enzymatic steps. Manipulation of these pathways by molecular biological techniques may be possible if the relevant genes can be isolated and transferred to a single plasmid which is then inserted into the genome of the desired host. Two groups of defensive proteins, lectins or haemagglutinins, and proteinase inhibitors may be more readily amenable to transfer to 'foreign' hosts, however. Lectins have highly specific sugar-binding properties and, in vertebrate herbivores, cause a number of effects, including agglutination of blood cells and stimulation of mitosis. There is also evidence for a defensive role against insects. Proteinase inhibitors are small, stable proteins that bind tightly to proteinases and appear to hinder digestion by herbivores, including insects.

As plants have evolved defences, herbivores have overcome them in order to survive. A plant diet presents two main problems for animals: that of overcoming defences and that of adapting to imbalance in the diet, usually a relative deficiency of nitrogen or other essential substances. Diet quality imposes on herbivores the constraints of identifying their correct food species and the most nutritious parts and growth stages of that plant.

Two avenues of adaptation to plant defence are evident. Some herbivores are generalists, depending on a wide range of food plants but at the expense of having adaptable digestive and detoxification systems.

16

However, most herbivores, including most insect pests, are specialists and have evolved adaptations that enable them to overcome the defences of the few plant species they attack. In many cases adaptation has progressed to the point where the herbivore depends on the characteristic defensive substances of the plant for identifying suitable food. Novel changes or additions to the secondary metabolite complement of a plant could disrupt these evolutionary adaptations.

Plants have defences that tend to be organised to maximise protection for a minimum metabolic cost. These defences render them virtually immune to all but a very few of the herbivorous species that encounter them. In addition, the individual plant can be more economical in allocating resources to defence when it possesses more than one kind of resistance. Variation in defence mechanisms exerts multiple selection pressures on the pest population, so preventing close adaptation. Generally, resistance of different kinds will be found in a crop if the germplasm is carefully searched, and effective defence may be achieved by breeding for a defence system with several resistance components.

As noted in Section 2.4.3, pests can be controlled by means other than exploiting genetic resistance. These aspects of pest control, which for insects include the exploitation of behaviour-controlling chemicals, are beyond the scope of this report.

2.6 Resistance to nematodes

Nematodes are potentially serious pests of wheat, oilseed rape and faba beans, among other crops. Although there are occasionally serious infestations in limited areas the average losses in yield caused by nematodes are believed to be very small. Many species can infect plants but the most damaging ones are those which are highly specialised, with limited host ranges. Like fungal pathogens and insects, the most damaging species exist in more than one race or pathotype and have major genes for virulence which are matched by major genes for resistance in host species.

2.7 General considerations on quality

For grain crops, quality is largely determined by genotype and the influence of environmental variation is usually comparatively small. In contrast, in the case of crops harvested for their vegetative organs (sugar beet, potatoes, etc.) the influence of environment on quality tends to be greater. In grain crops, the quality remains more or less stable after harvest and can be assessed unequivocally by chemical analysis. Many of the constituents of interest are present in high abundance though minor constituents with important effects, either desirable or otherwise, can be present. The major constituents of seeds are usually synthesised in the seed itself from intermediates imported from the mother plant. The synthetic machinery in dicots is determined by the genotype of the seed (i.e. it is influenced by the pollen as well as by the female parent) but is is regulated, at least in part, by supply factors determined by the female parent. The bulk of the seeds of oilseed rape and faba beans, like those of many dicots, consists of embryonic tissue. In wheat, as in other cereals, the embryo is only a small proportion of the grain, which mostly comprises triploid endosperm tissue, and the contribution of the maternal genes is twice as great as that of those from the pollen parent.

Quality can be defined only in relation to a given end use, and is usually a complex of features, some of which are more important than others. Major improvements in the quality of oilseed rape for human consumption and in the suitability of wheat for bread-making have been achieved recently by 'conventional' breeding and genetics, but further improvements are possible and, given continued investment in research, will be achieved. With faba beans, at present only a minor crop in the EEC, the end use will mainly be as a protein source for animal feeding-stuffs. No serious quality defects for this use have been identified, though if faba beans were to be a significant ingredient, feed compounders might discern the need for modified seed quality.

3

Molecular biology and plant breeding

In the context of this report, molecular biology is the study of genes (DNA), their primary products (messenger RNAs) and their secondary products (proteins). Molecular biology thus complements 'classical' plant biochemistry, which deals with tertiary gene products, that is, molecules made or transformed by enzymes, including intermediary metabolites, secondary metabolites and those which have a structural and storage role.

The upsurge of research in molecular biology in recent years has been made possible by the improvement in biochemical techniques that were first developed many years ago, the discovery of restriction endonucleases and the capacity to manipulate genes following the discovery of methods for recombining DNA molecules and propagating them in *E. coli*. More recently, there have been considerable improvements in the methods for determining the base sequences of DNA and for synthesising oligonucleotides semi-automatically. Also, it is now possible to insert genes into some plants so that the effects of specific modifications of a plant genotype can be determined. Together, these techniques will enable new insights to be gained into many hitherto unresolved questions in plant biology. Foremost examples are the molecular basis of responses to environmental factors and the regulation of differentiation and development. As noted earlier, this knowledge of fundamental processes will enhance assessment of the scope for engineering particular changes in plant form and function and so enable attainable targets for genetic manipulation to be defined. The techniques themselves will provide breeders with the means for making directed changes in plant genotypes, so creating new opportunities for plant improvement programmes.

The essential features of these techniques of molecular biology which can be exploited in plant breeding are described below. Their place in the entire cycle of plant breeding is illustrated in Fig. 3.1, but they also have applications for diagnosis and selection in 'conventional' breeding.

All plants contain three genomes, that of the nucleus, which contains by far the most DNA and most of the genes which are expressed, that of the chloroplasts, which codes for some 80–100 proteins, and that of the mitochondria, which codes for 20–30 proteins. Most of the proteins in the organelles are coded for by nuclear genes, however. Chloroplast and mitochondrial genes code for products which are constituents of the photosynthetic and respiratory pathways. Many instances are known of uniparental inheritance, implying that variation in non-nuclear genes also governs physiologically important functions other than photosynthesis and respiration, such as resistance to certain diseases and life-cycle duration. Prospects for manipulation of these non-nuclear genomes are considered in Section 3.5.

3.1   Gene isolation

It is now possible to isolate single genes from any organism. This is achieved by cloning DNA fragments in either plasmid, cosmid or bacteriophage vectors. Many complete libraries have been constructed from whole plant genomes and numerous genes have now been isolated and completely sequenced. This technology is established in many Euro-

Fig. 3.1 The breeding cycle showing conventional ——, new — — and proposed genetic engineering — · — methods of gene transfer and evaluation.

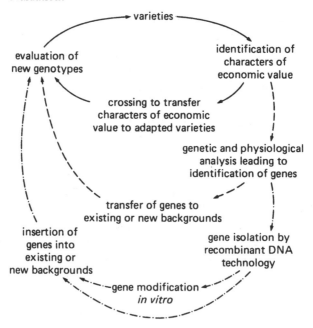

pean laboratories but it needs to be transferred to many others, particularly those concerned with understanding plant biology and with plant improvement by breeding. Although it is possible to purify single genes by cloning in *E. coli*, to recognise the gene of interest it is necessary first to have a messenger RNA (mRNA) for that gene, a DNA copy of the mRNA (cDNA) and knowledge of how it is regulated or, alternatively, an antibody which has been made to its protein product. Thus considerable information about the function of a gene is needed before it is possible to isolate the gene. Fragments of DNA can now be cloned in vectors which lead to a plant protein (or part of it) being made in *E. coli* or other cells. This allows the cloned DNA fragments to be recognised by their protein products rather than by the presence of the foreign DNA in the bacterial cell. More recently, new vectors allowing individual proteins to be synthesised in vitro with commercially available *E. coli* RNA polymerase and reticulocyte extracts have been used. As a means of identifying genes by their products this may become more important than expression in bacteria and other cells.

There is the prospect that transposable elements may provide a valuable technique for the identification and isolation of genes. Transposable elements are short segments of DNA with terminal inverted repeats flanking a sequence which may code for a protein or proteins which can excise the element at its ends and ligate the element back into a chromosome at a different site. Where this receptor site is within a gene, the gene will be inactivated. There is evidence that plant genomes contain very many repeated sequences dispersed throughout their genomes and this is taken as evidence that during evolution transposition has occurred on many occasions, though many of the transported sequences have lost their mobility and become 'fossilised'. Active elements are known in *Zea* and *Antirrhinum* (Fincham and Sastry, 1974) and if such elements could be detected in other crop plants or transferred to these from *Zea* or *Antirrhinum* they could be used for gene identification. The procedure would work best for the isolation of dominant genes present as single copies and would probably be as follows. An individual containing a transposable element would be crossed with another individual containing a particular phenotype of interest. Among the F_1 progeny, which would all be of one phenotype in the absence of a transposable element, some exceptions would be found which displayed the recessive phenotype. Some of these exceptions would have occurred because the transposable element had been inserted into the gene of interest so that it became inactive. These would be detected by probing with the DNA of the element. From a plant containing the inactive gene a genomic library would be prepared and the element in its new location with flanking sequences (i.e. parts of the gene of interest) identified. Probing the DNA of a normal plant with these flanking sequences would then reveal the intact, dominant allele of the gene.

3.2   Detection of genes and gene products

The availability of specific pieces of genetic information (i.e. DNA) from an organism makes is possible to detect that or a similar sequence of DNA in other individuals of the same species or of different species. The application of restriction endonucleases to isolated DNA before hybridisation (probing) of that DNA with the isolated DNA sequences enables allelic differences within and between species to be recognised, along with the copy number of the genes in any individual. This allows the DNA of individual plants to be screened and allelic differences to be recognised at the gene level.

The purified nucleic acid probes also enable the concentration of particular mRNAs to be determined in specific plant tissues or following environmental changes. For example, it is now possible to determine the response of a plant to heat stress by detecting the production of particular mRNAs, by hybridisation of total mRNA with copies of genes expressed only in response to heat. With these approaches it is possible to determine which genes respond to a change in environment or respond at different stages of development in terms of the production of their primary product, mRNA, and thereby define better the genetic basis of differentiation and development.

Genes which are activated during development have now been isolated from numerous crop species. Examples are those which are activated during seed formation, or which respond to light, for example the proteins involved in photosynthesis, or respond to stress conditions, for example heat-shock proteins, or which respond to symbiotic associations, for example leg-haemaglobin, or which are formed in response to invasion by pathogens.

The isolation of DNA sequences from plant chromosomes makes available many sequences which may be unique to particular species or variants of species. These sequences can be used as probes to detect the presence of the sequences in hybrids. These hybrids may have been produced by crossing within a species, by interspecific or intergeneric crossing or by DNA insertion. These probe techniques are often, but not always, more convenient than the techniques conventionally used in plant breeding programmes to recognise alien DNA.

The technique of using a purified gene to recognise that gene in another species provides a convenient way of screening plant material for the presence of pathogens. DNA copies of the genes of the pathogen, be it a virus, viroid or fungus, can easily be cloned and used to detect the presence of the pathogen in plant populations. This is a widely applicable technique which could be extended to a large number of plants. It might be developed for the identification of particular pathogen races.

Individual proteins can be recognised very specifically by antibodies. Antibodies are therefore extremely useful for recognising the synthesis of new proteins in plant development. Methods for producing polyclonal and monoclonal antibodies and their use in enzyme-linked immunosorbent assays (ELISA) and radioimmunoassays (RIA) are now very well developed. They are also a useful means of recognising unknown proteins, with a view to purifying the proteins and discovering their function. For example, one approach for uncovering mechanisms of resistance and pathogenicity might be to select antibodies against surface or extruded proteins from virulent genotypes of a pathogen that interact or show no interaction with proteins of avirulent genotypes. Antibodies have great potential in plant research and are currently under-used.

When genes have been isolated they can be easily sequenced, thereby giving information on the amino acid sequence of the protein specified by the gene. In turn, this can enable the tertiary structure of the protein to be determined. The presence of introns in the gene can be established and some of the factors outside the gene which are responsible for its expression can be recognised and evaluated. However, much more work is necessary before these factors are fully understood.

Restriction fragment length polymorphism (RFLP) provides a means of identifying DNA sequences which are closely linked to genes of interest to breeders without these genes needing to be explicitly identified or isolated. The principle is to correlate the occurrence of one or more DNA sequences having a known mobility upon electrophoresis in a gel (i.e. sequences containing a specific number of bases) with the expression of a particular character of interest. The sequences recognised will usually contain a considerable proportion of non-coding DNA and it is the polymorphism in this, rather than in the coding sequences that will often be the basis of the polymorphism. The first step is to establish a correlation between the character of interest and the presence of particular restriction fragments. This could be done by analysis of varieties either possessing or not possessing the character. Following this, the presence of these restriction fragments in individuals among segregating progenies indicates that it is highly likely that an individual plant in a segregating progeny possesses the character. The RFLP technique thus affords a means of improving the efficiency of selection which can be done on plants before they reach maturity, obviating the need to grow all individuals to a stage where the character is expressed. For the major crops, especially wheat, it may be desirable to identify several restriction fragments for each chromosome, distributed over its length, so that a set of 'ready-made' markers is available. It would then be easy to establish for any given gene to which fragment or fragments it was closely linked. Following this, selection for the fragment would generally equate to selection for the gene of interest.

3.3   Construction of new genes

It is possible to change an isolated gene at specific places in its base sequence to cause changes either in the protein it codes for and/or in gene expression. The ability to join pieces of DNA together *in vitro* before insertion into another organism enables hybrid genes to be constructed. These hybrid genes can specify the production of new proteins or of proteins that are expressed at different levels or in different tissues. This possibility clearly has the potential to change the production of enzymes during plant development and perhaps radically alter plant development in a way that has not been observed in nature. Similarly, it is possible to alter the location of an enzyme in a cell by joining the genetic information for the enzyme to that for a small peptide which causes the enzyme to be located in a specific cell compartment. For example, a bacterial enzyme has been inserted into the chloroplast of a higher plant by making such a hybrid gene. The effects of precisely modified and reconstructed genes on plant phenotype and performance can be tested. This technology will become especially valuable in plant breeding when it is known at a molecular level what changes are needed. For example, experiments are already beginning to produce storage proteins with changed amino acid balances in the hope that such changes will lead to crops with improved quality without loss of yield.

3.4   Insertion of genes into plants

A key step in the genetic engineer's technology is the ability to insert genes into plants. This is now possible for several dicot species using the pathogen *Agrobacterium tumefaciens* and there has been considerable progress in this area within the EEC programme during the last five years. With this technology it is possible not only to introduce different genes as indicated above but also to alter the number of copies

24

of existing genes or the position of a gene in the chromosome. These changes may lead to variation that has profound effects on the phenotype which have not been assessed by plant breeders because the variation does not exist in nature. In order for such altered genes to be expressed during development in specific ways or in response to specific environmental changes, it is necessary to identify regulatory genes in the recipient species, or, if they are absent, to introduce them. Much work is currently in progress to identify these regulatory genes and their essential sequences.

Although *A. tumefaciens* can be used as a vector in many dicotyledonous species, it is not a pathogen of monocotyledonous species, which include the *Gramineae*. There is an urgent need, therefore, to discover means for inserting genes into cells of monocots and regenerating plants from these cells, so that genes can be inserted into crop species of major importance for man, notably wheat, rice, maize and other cereals. Recent publications suggest that *Agrobacterium* could be used for transferring DNA to chromosomes of the monocots. This and regeneration are areas of research which must be followed up if cereals are to be improved by molecular biological techniques.

Although it is a major achievement in plant science to be able to insert into a plant any gene or variant of a gene from any other species, it seems likely that only one or a limited number of genes can be inserted at any one time, and for the most effective programmes these genes must have been recognised, purified and treated in a defined way before insertion. Therefore the contribution of gene insertion to plant improvement will proceed by steps, each step involving only one or a small number of genes. However, a single gene may have a large effect on the phenotype. For example, major genes affecting hormone biology may have a dramatic effect on plant phenotype. Similarly, one or a few genes affecting susceptibility to disease or the quality of the harvested product could have a major impact on the value of the crop. The technology of gene insertion thus provides opportunities for new strategies, for example in controlling diseases, and in the use of herbicides etc.

The advantages of being able to insert single genes or a small number of genes directly into a cultivar may eliminate the need for backcrossing (the traditional method of gene transfer), assuming that the genes are available to the plant breeder within the same species. Clearly, if the gene is not available in the species or has been modified in specific ways *in vitro* then the advantage is not only one of time in avoiding backcrossing and linkage to deleterious genes but also one of providing genes not previously available to the plant breeder.

3.5   The chloroplast and mitochondrial genomes

The application of recombinant DNA techniques to the manipulation of chloroplast and mitochondrial genes presents special problems because there are usually many chloroplasts and mitochondria per cell. Whereas the insertion of foreign DNA into the nuclear genome of a cell will usually ensure its inheritance in cell lines derived from that cell, no means of simultaneously transforming all the chloroplasts or mitochondria in a cell are known, and a single modified organelle would need to be positively selected if it were to survive and replace the original forms. Means of achieving this by inserting, along with a gene of interest, sensitivity to an antibiotic have been devised by de Block *et al.* (1985). When a cell containing a transformed chloroplast is treated with the antibiotic the normal chloroplasts are killed and the surviving chloroplast containing the gene of interest plus antibiotic resistance 'takes over'. This procedure requires that the antibiotic resistance does not affect fitness in the absence of the antibiotic, and that the antibiotic acts primarily on the chloroplast.

Protoplast fusion, giving parasexual hybrids, may offer means of making changes to the non-nuclear genomes of plant cells. Up to the present, parasexual hybrids have been obtained mainly by fusions within three families, the Solanaceae, the Cruciferae and the Umbelliferae. A few mitotic cycles after fusion, the nucleus may be from one or other donor. There may be a mixed population of chloroplasts but, as far as is known, none which are genetic recombinants. On the other hand, the populations of mitochondria in a cell may contain both parental types and recombinants. Thus there is an opportunity to select (at the level of either the cell, or the tissue or the regenerated plant) for either chloroplast or mitochondrial type or for a particular recombinant mitochondrial genome, provided a suitable selection system exists or can be created.

3.6   Molecular biology and plant physiology

There remain many areas of plant physiology and biochemistry where the molecular basis of processes, the control and regulation of these processes and their sensitivity to environmental factors is very imperfectly understood. For example, one of the most intensively studied processes, photosynthesis, is well understood in terms of its subcellular organisation, the main features of the light harvesting machinery and photosynthetic electron transport, and the pathways of photosynthetic carbon metabolism and of sucrose synthesis, but despite this, there is no general agreement on whether photosynthesis limits the growth of crops and, if it does, what particular step or steps are limiting. From what is known, numerous questions remain unanswered on quantitative control and regulation mechanisms. Some of the more important questions are as follows. What is the mechanism which appears to regulate within narrow limits the internal carbon dioxide concentration in the intercellular spaces within leaves? What is the mechanism of carbon dioxide transport from the surfaces of mesophyll cells to the chloroplasts and what are the components of the resistance involved? How important for quantum efficiency are the mechanisms, which have short relaxation times, for regulating the distribution of the light intercepted by the leaf to the two photosystems? What is the mechanism for regulating the stoichiometry between the light harvesting machinery and the complement of Calvin cycle apparatus? What is the nature of the photoinhibition of photosynthesis? Does the amount of ribulose bisphosphate carboxylase/oxygenase limit carbon dioxide fixation per unit leaf area? How is the amount of the enzyme regulated, particularly by the carbon dioxide concentration? Uncertainty over these and other aspects of the photosynthetic apparatus makes it impossible for physiologists to identify targets for genetic manipulation which would guarantee success in increasing photosynthesis and ultimately yield. At the organ and whole plant level, while it is well known that there are differences among ecotypes of a species or cultivars of a crop in the photosynthetic characteristics of single leaves, there is no unequivocal evidence that, among genotypes, high rates of photosynthesis per unit leaf area are correlated with high growth rates or with plant yield. Indeed, there are usually correlated changes in leaf size and leaf thickness (or some other measure of the amount of photosynthetic apparatus under unit leaf area) which compensate for high photosynthetic rate such that, within a species, ecotypes or cultivars have remarkably similar growth rates. It is presumed that the correlated changes confer adaptation of ecotypes or cultivars to particular environments, differing in their water availability, light intensity, etc. Thus, determining whether it is possible to increase plant growth rates through increasing photosynthetic rates by genetic means remains a major unsolved problem in plant biology. Clearly, a concerted effort on a broad front is needed to resolve the issue. It may be expected that, through molecular biology, the ability to make small, defined changes in putative key regulatory molecules will provide a powerful means of advancing knowledge in this area.

What is true of photosynthesis is also true for differentiation and development. Despite much research, the roles of plant hormones at the molecular and cellular level are hardly understood at all. How is the production of these substances regulated by the environment? In which cells are they produced? How are they transported? What determines whether a cell is a target for hormone action – even, do hormones play the regulatory role that much classical plant physiology has supposed? All these are largely unresolved questions. Although it is now known that hormones can affect the transcription of some genes, it seems unlikely that control at this level would provide an adequate model for hormone action in total. Other major areas of uncertainty are the control of senescence, the remobilisation of materials and the responses to stress. Molecular biology is providing a powerful new means for investigating all these aspects of plant function and, in conjunction with well-established methods in physiology, biophysics and biochemistry, should enable major advances to be made. However, it is at present inconceivable that our knowledge will ever be adequate enough to predict with certainty the consequence for yield, quality or disease resistance of a given transformation, and so this will have to be determined empirically. This evaluation is likely to be a multi-stage process involving measurements at different levels of organisation and culminating in yield trials (carried out over years and locations) comparing the control and the transformed plant. Only when beneficial effects have been demonstrated at this level will breeders assume an active interest. To be detectable in field experiments, transformed plants would have to yield at least 3–5 per cent more than the control in an appropriate range of environments. It follows that only those transformations considered likely to give increases in yield of 3–5 per cent or more are likely to be worthwhile. Similar considerations apply to changes in disease resistance and quality.

3.7   Molecular biology and resistance to disease

There is a need both for (a) research directed towards greater understanding of resistance gene action and the related molecular mechanisms, and for (b) research aimed directly at improving disease resistance of crop plants. The latter involves the need to (i) identify sources of disease resistance, (ii) transfer resistance genes, and (iii) monitor the presence of resistance genes and evaluate their activity. In general, all of these objectives may be achieved by conventional, current methods of testing, breeding and selection, but with varying degrees of success and often only following prolonged effort.

Applications of molecular biological methods to disease problems could include:

1. the elucidation of mechanisms of resistance

2. the identification and characterisation of resistance genes and their protein products
3. the identification and characterisation of avirulence genes, their products and alternative functions
4. the modification of resistance genes in order to confer greater effectiveness or stability
5. the development of probes to detect resistance genes, so allowing rapid screening of progenies, surveys of germplasm, etc.
6. the transfer of resistance genes both within, and perhaps more importantly, between species
7. the genetic modification of organisms antagonistic to pathogens, so as to effect 'biological control'.

There is a particular need to understand the nature of resistance which proves 'durable', that is able to persist for many years and not be overcome readily by rapidly evolving pathogen populations.

The achievement of many of the above objectives will probably depend initially on the success of largely random searches to isolate a resistance gene or its protein product. The resistance gene product is likely to be a very minor component of total plant protein and its isolation will depend on the use of appropriate 'high resolution' techniques of analysis which exploit specific distinguishing properties. Protocols have been suggested involving analysis of plant plasma membrane (PM) fractions. One possibility would be to raise antibodies to crude PM preparations from a resistant cultivar and react these against a similar preparation from an isogenic line lacking the resistance gene. The aim would be to isolate antibodies which bind uniquely to a protein component from the resistant cultivar. The subsequent purification of receptor protein could then be achieved by affinity chromatography. Isolation of an antibody to the receptor protein would permit recognition of the protein's mRNA during *in vitro* protein synthesis, and could lead to the subsequent isolation of the resistance gene by established methods of molecular biology. More 'direct' approaches towards isolating disease resistance genes may be feasible. Some techniques may, however, be applicable only in restricted circumstances (e.g. the use of transposable elements in maize). Most involve some element of chance. The technique of cascade-hybridisation enables tissue-specific mRNAs to be isolated and would be suitable for isolating resistance gene mRNA providing that lines isogenic for the gene are available and that the resistance gene is expressed at a sufficiently high level. Basically, this technique involves hybridisation of mRNA isolated from a resistant isoline with DNA from a susceptible line. The mRNA which fails to bind represents sequences unique to the resistance gene. From this, cDNA probes could be constructed and used to detect resistance gene sequences.

Having once located and sequenced one resistance gene, common sequence features likely to be present in other, related genes would also allow their rapid location and isolation. Resistance gene probes, once available, could be used for a number of purposes including screening of germplasm for sources of resistance and selection of progeny into which resistance genes may have been transferred. Direct probing would, in principle, eliminate the need for empirical testing by exposure to the pathogen; however, it would still be desirable to check that the gene, when present, was expressed.

Thus far, attention has been focussed mainly on genetic manipulation and 'probing' of host genes. Similar effort should be given to investigation of the pathogen. In particular, the possibilities of engineering organisms to effect biological control of pathogenic species might be considered. These could act either externally, by competing with, or antagonising the development of the pathogen during stages prior to infection of the plant, or internally in the form of viral infection. The cytoplasmic hypovirulence factors associated with loss of virulence of otherwise extremely pathogenic fungi and which consist of double-stranded RNA, would appear to be strong candidates for molecular analysis and manipulation. Certain bacteria are known to inhibit development of the take-all fungus disease of cereals. More active strains of these might be developed. More generally, fungal or bacterial strains might be 'engineered' to produce high amounts of antifungal compounds such as those which prevent rust spore germination on the leaf surface.

The molecular biology of plant viral infections is relatively well developed compared to that of fungal pathogens. Pathogenesis-related (PR) proteins have been isolated following viral infection of tobacco, and attempts to clone the genes coding for these are being made. These proteins are inducible and can be elicited by treatments other than virus infection, but their precise function is unclear. It is also known that a prior virus infection can afford some protection against subsequent infections. Insertion of viral sequences providing such 'immunity' into plant genomes is thus an attractive possibility.

3.8   Molecular biology and resistance to pests

Before molecular biology can contribute to the production of useful pest-resistant crop varieties, it is necessary to identify genes that will confer resistance and to select genes that are likely to repay the effort and cost of transfer. There seems little chance in the foreseeable future of transferring the genes for complex biosynthetic pathways or a number of unrelated genes governing differing, complementary resistance mechanisms. Advances sought by molecular biological methods should involve transfer of single genes, at least initially. The transferred genes could increase, alter or add to the normal defences of the transformed plant.

Much has been made of the supposed contrast between resistance due to single genes, which is held to be easily overcome by specifically-adapted parasite genotypes, and polygenic resistance, which is regarded as permanent. In practice, there are many exceptions and intermediate examples. Prediction of probable durability from this simple contrast is likely to be unproductive in evaluating the prospects for introducing durable resistance into crop varieties. Single genes have given long-lasting resistance to pests, and resistance generated by transfer of a single gene will not necessarily fail because only one gene is involved.

Broadly, two groups of genes may be considered. The first comprise those, probably from related plants, that cannot be transferred easily by conventional methods but are likely to operate in conjunction with the biochemistry of the recipient. Importantly, there is the possibility of transferring genes for enzymes that could produce alien secondary metabolites by a single-step reaction from the normal intermediary metabolites of the recipient. Genes in the second group may come from a genome unrelated to that of the recipient and lead directly to alien products within the recipient. Such genes may be looked for not only in plants but also in animals or any other organism, or might be generated by *in vitro* mutation, or constructed *de novo*.

The crystalline protein toxin (delta endotoxin) produced by *Bacillus thuringiensis* has been intensively studied as a potential insecticide. More is known of this material than any comparable potential plant-defence compound. Toxicity to insects is due to polypeptides representing half the molecule of the crystalline protein. Serotype variants of delta endotoxin, from different varieties of *B. thuringiensis*, differ in specificity, affecting some groups of insects more than others. Lepidoptera tend to be most susceptible to most of the known serotypes, but a second group of endotoxins is most toxic to Diptera. The delta endotoxin gene occurs in a plasmid and recently, successful transfer of a *B. thuringiensis* gene to tobacco has been reported, with toxin being present in the transformed cells.

Similarly, other microorganisms and viruses that are pathogenic to insects are possible sources of potential resistance genes. Such resistance genes, once characterised, cloned and associated with vectors, could probably be used in a very wide range of crop species if their function was independent of the crop genome. There is, however, the chance that the uncontrolled synthesis of resistance factors, especially proteins, might divert the plants' energy and carbon resources and reduce yield.

A major block to progress in transferring natural resistance mechanisms is the lack of knowledge of their function at the gene-product level. Causes, or at least contributory causes, of resistance to insects are known in several cases, but with a few exceptions, little is known of their inheritance, or where secondary metabolites are involved, or of the enzymes governing their production. The genetics of resistance to some fungal pathogens is better known but again, at the gene-product level, causes of resistance are usually unknown. Without this information it is difficult to identify the genes.

Among plant-defensive chemicals, proteins, including lectins and proteinase inhibitors, stand out as being most likely to be easily handled by molecular biological methods. More needs to be known about the protein inhibitor inducing factors (PIIFs) which mediate increases in resistance in wounded plants, and about the role of lectins in resistance to insects.

As already noted, insects of one species do not feed on the great majority of the plants that they encounter. Although insects respond to the overall 'balance' of chemical stimuli from their food plants, in most cases non-adapted insects are probably prevented from feeding by a single substance, typically a secondary metabolite. Thus, it seems likely that introduction of an alien substance to a crop could give resistance to many of the pests that normally attack it.

Most secondary metabolites are formed via a complex biosynthetic pathway, so that transfer of genes causing their production will also be complex (but see Section 2.5). However, it can be envisaged that single genes could be identified and transferred to divert the intermediary metabolism of the recipient to generate novel and alien substances. Addition of an alien defensive 'flavour' to a crop's natural defence could confer protection against both the usual, adapted pests and the pests of the donor. There is a risk, however, that introduction of biosynthetic machinery for secondary metabolites into unadapted plant genotypes will cause autotoxicity, or adversely affect crop quality.

Microorganisms play roles of varying importance in the nutrition of herbivores, but are often essential as their activities remedy deficiencies in diets of plant tissue or sap. Introduction to a crop of the ability to synthesise antibiotics to destroy symbiotic microorganisms could be a useful way of generating resistance to insect multiplication.

32

Whilst it is clear that molecular biological techniques may offer the opportunity of short-cutting conventional plant breeding methods by introducing novel or inaccessible variation to crops, the effort applied to a particular pest problem needs to be assessed carefully. Because insect populations are held in check by many agents, including predators, parasites and the weather, in addition to plant defences, pest resistance need not be very conspicuous to have a substantial effect. Natural resistance to many pests is likely to be easily accessible to plant breeders at levels that could prevent significant damage in all but rare circumstances. It is likely that screening to select among breeding stocks for existing variation in resistance would suffice to produce adequate levels of resistance, if suitable methods can be developed.

3.9 Molecular biology and crop quality

The crops considered in this report are harvested for their seed, and in all of them the seed proteins are of economic importance. Major advances are now being made in elucidating the primary and secondary structures of seed storage proteins and in determining the extent of homology between these proteins from diverse seed plants. The new knowledge will provide a scientific basis for assessing what changes in these proteins are biologically feasible. For example, there is a substantial variable region in the 11S storage proteins in legumes and other plant families. It should be possible to increase the abundance of methionine in this region (though not in other regions), so improving the nutritional value of the proteins. This could be done either by construction and insertion of DNA sequences or by transfer of desirable sequences present in other plants to the crop of interest.

Further examples are given in Sections 4.6.4, 5.5.4 and 6.5.4.

3.10 Molecular biology and plant cell culture

The techniques of plant cell culture are likely to continue to be of key importance in the application of molecular biology to crop improvement. It is beyond the scope of this report to consider the techniques in any detail, but aspects of them which are at present limiting are noted below.

The culture of undifferentiated plant cells, either aggregated into calluses or singly was first achieved many years ago and is now widely used as a means of propagating plants. However, in culture media and freed from an environment of differentiated tissue, mitosis is frequently abnormal and, with increasing numbers of divisions, calluses can contain many cells which are aneuploid. In addition, loss of parts of chromosomes can occur. Some of the abnormal cells can survive in callus cultures but have lost the capacity to regenerate into entire, fertile plants. Others do produce entire, fertile plants but have an altered phenotype. This variation has been regarded as a potentially valuable means of making genetic changes in crop plants to improve particular features in otherwise desirable varieties, but there is at present insufficient experience to assess whether it will be a useful complement to conventional breeding.

Clearly, in micro-propagation, genetic change should be minimised, and this is achieved by minimising the exposure to potential mutagens, including plant hormones, and reducing or eliminating the phase of undifferentiated cell division.

Plant species, and varieties of the same species, differ greatly in the ease with which they can be induced to form callus cultures. Likewise there is much variation in growth rates and in the regeneration of whole plants from calluses. Species which can easily be induced to form calluses which will readily regenerate include many Solanaceae, some Leguminosae, some Brassicas, carrots and asparagus, while little success has been achieved with the Gramineae. The reasons for these differences are not understood and so at present progress is by trial and error and is expensive and time consuming.

The best known transformation system which is effective in higher plants employs *Agrobacterium tumefaciens* and callus cultures. There is a need for improved and more predictable callus culture methods which minimise chromosomal damage. The need for a transformation system for cereals has already been noted (Section 3.4). Other transformation systems, such as those based on virus vectors and naked DNA with appropriate selectable markers, need to be further developed and tested.

Free cell cultures offer the prospect for mass selection for biochemical mutants. Provided that the relevant genes are or can be induced to be expressed, mutagens and appropriate selection systems can be used to isolate mutants which occur at very low frequency, by techniques analogous to those used for fungi and bacteria. This is potentially a very powerful method for making defined genetic changes but the critical step is to devise a selection system appropriate for detecting a desired genetic change.

Protoplast fusion offers a means of 'hybridising' plants where sexual hybridisation is not possible and could have value in the transfer of genes from other *Vicia* species to faba beans. While the nature of the hybridisation that takes place when protoplasts fuse is not known, there is the interesting and potentially important possibility that chloroplast and mitochondrial, as well as nuclear hybridisation may occur. Further, there is the possibility that genes for complex characters could be trans-

34

ferred even if the genes and their locations are not known, but again, as transfer may be at very low frequency, it would be very desirable to be able to select at an early stage after fusion. It has been reported that this technique has been used successfully to construct a male sterile, atrazine resistant *Brassica* plant (Pelletier *et al.*, 1982).

4
Wheat

The economic incentives offered by the EEC Common Agricultural Policy, coupled with the availability of new technology, have greatly stimulated the production of wheat in member states both in terms of the area grown and the yield of grain per hectare. Taking the UK, the wheat area has increased from 0.8 M ha in 1970 to 1.96 M ha in 1984, and the yield per hectare from close to 4 t ha^{-1} to 7.5 t ha $^{-1}$ in 1984. Similar increases in yield have occurred in France, Germany, Holland and Belgium. The EEC is now a net exporter of wheat and is in competition with lower-cost producers. Clearly, in breeding new varieties emphasis needs to be given to improved quality to make the product more attractive to purchasers. To reduce production costs, increased resistance to pests and diseases is needed. However, as long as land prices remain at their present level in real terms, farmers will also be able to reduce unit costs of production as or more effectively by growing higher yielding varieties. Thus there will remain the incentive for breeders to produce higher yielding varieties. In summary, therefore, breeders' broad objectives are not likely to be influenced substantially by the level of production of wheat in the countries of the EEC.

Numerous non-traditional uses for wheat grain and for wheat straw have been suggested. These outlets will gain acceptance in the long term only if they are economic in comparison with competing products. It is beyond the scope of this report to assess these issues. Instead, some non-traditional uses are listed below:

Non-traditional uses of wheat grain, chiefly starch

Paper industry	– for paper coating and surface sizing
Textile industry	– sizing for warp yarns; finishing; textile printing

Synthetic polymer industry	– biopolymers, graft polymers, fillers, adhesives
Biochemical industry	– enzymes, hormones, antibiotics, vaccines, by fermentation processes using genetically engineered organisms
Chemical industry	– Organic acids and solvents, including ethanol

Non-traditional uses of straw

Cellulose	– paper, fibreboard, biodegradable plastics, absorbants, fire retardants, binders, coatings, thickeners and via glucose, sweeteners, pharmaceuticals, organic solvents, alcohols, organic acids, etc.
Pentosans	– xylose, sweeteners, furfural and furan derivatives
Lignin	– energy, organic chemicals

4.1 Biological limitations to yield

Wheat is well adapted to the climate of north-western Europe. The generally moderate temperatures allow the crop a growing season of 10–11 months. In contrast, in parts of the US and the northern areas of North America cold winters dictate that wheat is sown in the spring, giving a cropping season as short as three months, whilst in lower latitudes the season is limited by high temperature and drought and is typically of six months duration.

As a framework for analysing limitations to yield, the life cycle of the crop may be divided into two phases. The first lasts from sowing until pollination. During this phase the yield potential of the crop is determined. The yield potential is defined as the number of fertile florets per unit area of cropped land. During the second phase, from pollination to maturity, environmental conditions determine the extent to which this potential is achieved, in terms of the percentage of fertile florets which sets grain, and in terms of mean weight per grain. It has been established in several studies that, except when there is a moderate to severe drought during grain filling, about 90 per cent of the grain carhohydrate originates from photosynthesis during grain filling. In contrast, most of the protein present in the grain at maturity is already present in the leaves and stems of the plant at the time of pollination and is progressively translocated, in the form of amino acids, from these organs as they senesce.

Yield potential is increased by any environmental condition or combination of genes that results in the production and survival of more tillers and, within the ear of each tiller, of more fertile florets. For reasons which are not fully understood, though usually rationalised as the consequence of intra-plant competition, there are pronounced negative correlations between tiller production and survival: between the number of surviving tillers and the number of florets per ear and between the number of florets per ear and potential grain size. This is generally true when genotypes varying in these features are compared and when they are varied by some factors of the environment. Progress in raising yield potential depends on discovering genotypes which are exceptional in that they can produce an above-average number of grains per ear for a given number of ears, or vice versa, or both. Thus modern wheat varieties generally produce more ear-bearing shoots and more grains per unit area of land than their predecessors (Table 4.1).

A high yield potential, as defined above, is both associated with and caused by the development of a high leaf area index. In turn, those leaves which persist until the grain-filling stage, together with other green surfaces, provide the photosynthetic capacity for carbon dioxide fixation and hence the supply of carbohydrate to the grains.

In Western Europe, increased use of nitrogenous fertilisers is the single most important factor which has raised both the yield potential and the realised yield. Extra nitrogen relieves the limitation on plant growth and leaf area development which is imposed by the slow rate of release of mineral nitrogen from soil organic matter and from the small amount of nitrate deposited with rain. To take advantage of the benefits from fertiliser nitrogen, breeders have progressively reduced the height of varieties. As shown by the results in Table 4.1, reduced height is also associated with increased yield. A major reason for this appears to be reduced competition from the growing stems for a limited supply of carbohydrate, during the time when both stems and ears are growing rapidly. As plant height has been reduced, so the density of leaves within the crop canopy (m^2 of leaf surface per m^3 of crop canopy) has increased. To maximise crop photosynthesis, it is necessary for the crop to intercept the maximum proportion of the incident solar radiation, and for the radiation to be distributed as uniformly as possible over the green leaves

38

and other photosynthetically active surfaces. With increased canopy density, light distribution becomes more critical, and requires the plants to have erect leaves.

For wheat, which is dependent on fertiliser nitrogen (because it lacks the capacity to fix atmospheric nitrogen), increases in yield in the future will depend on the use of more nitrogenous fertilisers, coupled with the production of varieties capable of bearing greater numbers of larger ears, and having a photosynthetic capacity and duration of grain-filling period that enables more, larger grains to be produced.

In the gene pool readily available to breeders, there is much variation for environmentally regulated developmental responses. At present breeders rely on the random reassortment of genes which produce minor changes in developmental patterns and, by selection for high yield there is a slow 'drift' towards improved yield potential. For example, some genes which cause earlier flowering, by reducing the time available for the development of yield potential, may lower yield potential. However, modern British varieties of wheat flower at least a week earlier than those grown in the nineteenth century. Though breeders have not consciously selected for earlier flowering, presumably earlier flowering is of benefit (e.g. by aiding drought avoidance) provided that its adverse effects can be compensated for by selection for genes which increase the rate of development. Recent experimental evidence suggests that it will be advantageous to produce varieties which flower even earlier than those currently grown. Progress in breeding for increased yield potential may become limited by lack of knowledge of the genetic control of development and flowering. Numerous genes are likely to be involved and to date only a few loci, governing response to day length and to low, vernalising, temperatures have been identified. Even for these,

Table 4.1 *Yields, yield components and heights of UK wheat varieties. Unpublished results from an experiment at Trumpington, UK, 1984/85*

Variety group	Grain yield (t ha^{-1})	No. of ears (m^{-2})	No. of grains (m^{-2})	Plant height to base of ears (cm)
Very old (pre 1900)	6.4	386	15 020	158
Old (1900–20)	7.3	392	17 340	141
Intermediate (1950–75)	9.2	432	17 970	104
Modern (1980–present)	11.3	457	23 640	85

there is no rapid means of determining the allelic composition of individual plants. Superimposed on the uncertainties of the genetic control of development is the inherent variability of the climate. There is no practical advantage in attempting to produce a variety optimally adapted to a fixed environment (even if this were possible) for, in real life, the environment varies with site and year. As far as can be assessed at present, while an increased understanding of the genetics and physiology of the control of development could help in the more rational 'design' of varieties, the acquisition and application of the necessary information might be more costly than justified when compared with the progress that could be made by direct selection for yield in the relevant environments. At present, we do not have the knowledge to assess whether the 'rational' or empirical approach would be the more cost effective. Not enough is known about the relevant genes and their effects, and we do not yet know how easy it will be in the future to introduce particular genes into new genetic backgrounds.

The attainment of a high yield potential and the realisation of a high yield depend on the crop having a sustained, rapid rate of photosynthesis as well as appropriate developmental responses to the environment. It has already been noted that crop photosynthesis is a function of leaf area and the rate of photosynthesis per unit leaf area, and that leaf area development is greatly promoted by fertiliser nitrogen. Nitrogen also reduces the rate of senescence of leaves, so helping to sustain a high leaf area. When nitrogen is not limiting, shortage of water and the photosynthetic capacity per unit leaf area impose limits on the rate of photosynthesis of the entire crop.

Wheat, though better adapted for water-limited environments than many crops, often experiences episodes of stress which reduce crop photosynthesis and yield. Water stress has been estimated to reduce wheat yields in England by an average (over years) of between 10 and 20 per cent. The yield loss in most other EEC countries is likely to be greater. When it occurs before pollination, water stress reduces photosynthesis, the rate of expansion and the final sizes of leaves, accelerates their senescence, increases tiller death and reduces the ear size. These responses are of adaptive value for the crop if there is a continued drought, for they reduce its rate of water consumption so that more is left for the grain-filling stage. Drought during grain-filling also reduces the rate of photosynthesis and accelerates leaf senescence, so reducing grain size. If it occurs early enough after pollination it can also reduce the number of grains which develop to maturity. Crop morphology can affect water use and yield response to drought. Thus it is possible to

40

develop genotypes which are best suited for particular patterns and degrees of severity of drought. Selection for high yield in a particular region achieves this objective, though not very efficiently. There are, however, two limitations to the better exploitation of drought resistance. First, because long-term weather forecasting is very imperfect, it is not possible to predict seasons when drought is expected and to sow drought-resistant varieties. Second, because trialling procedures do not take drought into account, varieties are not explicitly produced for different soil types on which the likelihood of drought is different. The decision to breed varieties for particular soil types and areas differing in water availability depends on the perceived benefits, in terms of yield and costs of production, from having special-purpose varieties. At present there is insufficient data from which to assess the benefits.

High photosynthetic capacity per unit leaf area should benefit yield under a wide range of water availabilities. In water-limited environments, it should increase the amount of crop photosynthesis per unit of water transpired. In those environments where water is not limiting, it will, particularly if coupled with large leaf area and an appropriate leaf disposition, enable the maximum photosynthetic capacity of the crop to be raised.

Calculations have shown that the maximum yield of wheat (10% protein, 15% moisture content) attainable in England when water and nutrients are not limiting, pests and diseases are absent and there is no lodging is between 12 and 14 t ha^{-1} for varieties currently grown. Yields obtained on favourable soils and in good years are 11–12 t ha^{-1}. Average yields on a country basis, 7.5 t ha^{-1} in the climatically favourable year of 1984, reflect the extent to which diseases, pests, drought and management factors limit yields. Clearly, even if management were perfect and pests and diseases completely controlled, yields would often be lower than the theoretical maximum because of drought and unfavourable temperatures.

Most of the wheat grown in the countries of the EEC is autumn-sown and varieties need to possess a degree of winter hardiness. European varieties do not need the extreme hardiness required in much of the USSR and Canada. The physiology and genetics of hardiness are imperfectly understood, and in particular it is not known whether there is a yield 'penalty' associated with hardiness. If there is, then selection for high yield will occur at the expense of hardiness and so, unless breeding material is regularly exposed to appropriately severe winters, varieties selected for high yield will be at risk in severe winters. There is some evidence for a role of the hormone abscisic acid in determining hardiness, and that a form of invertase is synthesised only during hardening at low temperatures.

Ways in which the maximum yield might be increased by breeding are summarised in Table 4.2. It must be emphasised that the table lists characters each of which is complex and, where known, controlled by several or many genes. Some of the characters can be selected for directly in the field and, in the long term, selection for high yield in a particular environment will produce close to optimal expression of the characters. Genetic and physiological analysis permits a more 'scientific' approach to breeding, but only if carried out on a substantial scale with relevant genetic backgrounds and in appropriate environments. The information and breeding material from such analyses should enable yields to be increased more rapidly by breeding than otherwise. Apart from cost, the disadvantage is that it is not practicable to take account of all the possible interactions between genes that may outweigh their main effects. From the standpoint of the application of molecular biology, progress will be possible only when particular genes with important effects are identified, implying the need for analysis of the genetic components determining the characters listed in the table. In turn this requires greater knowledge of the physiology, biochemistry and genetics of wheat. Some examples of opportunities for these approaches are given in Section 4.6.1.

4.2   Diseases

New varieties of wheat introduced into agriculture must possess satisfactory levels of resistance to major diseases such as yellow rust (*Puccinia striiformis*) and powdery mildew (*Erysiphe graminis* f.sp. *tritici*). Loss of resistance, due principally to changes in the nature of pathogen populations, is a major reason for the demise of varieties and necessitates continued breeding effort to replace them.

The introduction of disease resistance into a new variety is at present, as with other desired characters, generally achieved by conventional techniques of sexual hybridisation. As well as utilisation of variation within tetraploid wheat, appreciable use has been made of related species as sources of disease resistance. However, the process is slow and uncertain in outcome, and much potential resistance probably remains unexploited due to the technical problems involved.

Major diseases of wheat of importance in Europe are caused by yellow and brown rusts, powdery mildew, eyespots, take-all, *Septoria* spp. and *Fusarium*. Because of their wide distribution, economic importance, rapid modes of dispersal and infection, and accessibility for experimentation, rusts and mildews have been subjected to much more intensive

Table 4.2 *Crop characters and wheat yield*

Character	Advantage	Possible disadvantage due to pleiotropy
More tillers produced	yield component	smaller tillers and ears, increased sensitivity to pre-anthesis drought
More tillers survive	yield component	as above
Larger ears, more grains per ear	yield component	fewer ears
Larger grains	yield component	fewer grains per ear, increased sensitivity to late drought
Larger leaves	greater photosynthetic capacity of crop and nitrogen uptake	lower photosynthesis per unit leaf area and poorer distribution of intercepted light
More leaves produced	larger yield potential	delayed flowering, smaller grains
High rates of light-saturated photosynthesis	faster growth rates	more rapid senescence of leaves, smaller leaves
Lower rates of respiration as a proportion of photosynthesis	faster growth rates	not known
Longer grain-filling period	more grain carbohydrate	delayed maturity
Earlier flowering	drought avoidance	reduced yield potential
More erect leaves	improved light interception	increased sensitivity to a pre-anthesis drought
Higher assimilation ratio	drought resistance	reduced yield potential when water is not limiting
Larger, deeper root system	drought avoidance	not known
Presence of awns	greater contribution of ears to crop photosynthesis	increased sensitivity to ear diseases in humid climates
Shorter straw	reduced competition between developing stems and ears for carbon and nitrogen	reduced yield potential and nitrogen uptake
Greater nitrogen uptake	greater biomass and grain-protein yield	longer life cycle and delayed maturity

study than have many other wheat diseases, several of which are soil borne. Much of the evidence for race-specific, gene-for-gene resistance has been provided by the study of rust or mildew interactions, and very extensive information is available concerning the genetics of resistance to these pathogens, particularly in wheat. Monogenic dominant genes, some showing multiple allelism, often govern resistance, although the expression of resistance genes may be modified by the genetic background of the host, and some genes may suppress the expression of resistance genes. There are also some instances of recessive alleles conferring resistance. The best studied cases, however, are those where resistance is dominant, implying the active participation of a plant gene product in resisting fungal attack. Fungal avirulence genes are also generally dominant, and mutation by the pathogen to a greater degree of virulence is more likely than mutation to reduced virulence. (There are, however, exceptions to this; see below). Similarly, in the host, mutations which increase susceptibility may be more likely than those which increase resistance, hence the need to introduce specific resistance genes by directed breeding.

Although elicitor–receptor models currently dominate ideas on resistance mechanisms, there is little direct *biochemical* confirmation of these. No specific receptor proteins have been isolated, the nature of the hypothetical second messenger(s) remains unknown, and no phytoalexins have been chemically identified in *Triticum*, although there is some evidence for the participation of such compounds, and there are precedents for such substances in other cereals (oats, barley, rice). No host-selective fungal toxins capable of mimicking disease action have so far been isolated from infected wheat or its pathogens, though their presence has been indicated in experiments with brown rust and a non-selective toxin has been detected with *Septoria*. Toxins have been demonstrated to be products of other cereal pathogens.

4.2.1   Powdery mildew (Erysiphe graminis *f.sp.* tritici)

Considerable race-specific resistance to powdery mildew exists within *Triticum aestivum*. Most known sources of this resistance appear to have been exploited in breeding programmes and there is a need to utilise alien sources of resistance more freely. There is considerable variation for resistance in related species and genera including *Aegilops* and *Agropyron*. However, relatively few 'alien' resistance genes have been transferred into tetraploid and hexaploid wheat, probably because present techniques are laborious and uncertain. Also, past experience has shown that the resistance transferred is often not durable.

At least 11 distinct loci conferring mildew resistance now exist in different cultivars of *T. aestivum*. Host genes which confer susceptibility have also been found, and elimination or mutation of these would be one possible means of increasing resistance. This could prove more permanent than the introduction of a specific resistance gene which can be 'overcome' by new pathogen races. Such effects could be involved in the non-specific quantitative resistance present in some genetic backgrounds. More detailed genetic analysis to understand these effects and permit their greater exploitation has been recommended.

Individual race-specific resistance genes have been shown to act at specific stages of the infection process. (However, the gene products and the resistance mechanisms they initiate are still a mystery.) This has been achieved by carefully controlled experimentation involving the use of host isolines and of fungal isolates selected for a high efficiency and synchrony of spore germination. Such studies show that race-specific resistance genes operate only *after* penetration by the fungus into the host, when close contact between fungal and host cell membrane surfaces has probably been achieved. Prior to penetration, features of the host leaf surface may influence spore retention, germination and appressorial development and attachment. Rapid attachment becomes most critical at low atmospheric humidities, when the primary germ tube may rapidly exhaust the water supplied to it from the spore. Host features which might limit attachment or penetration are the degree of hairiness, cuticle composition and thickness, and wall silification in the leaf. Barley mutants with altered cuticular waxes are more resistant to infection by *E. graminis* f.sp. *hordei*, and malformed appressoria are produced by the fungus on these plants. It has been proposed that the normal wax layer stimulates the formation of mature appressoria. Whether this involves physical or chemical factors is not clear, though the importance of physical structure has been emphasised. Resistance to infection imparted by surface features is independent of that conferred by specific resistance genes. Clearly, there is scope for exploiting such features as a possible means of obtaining an increased level of general, non-specific resistance.

Penetration by *E. graminis* takes place through the cuticle and appears to involve the production of digestive enzymes. The 'clean' nature of penetration sites suggests that enzymes may be bound to the hyphal tip rather than released extensively into solution. Haustorial formation follows penetration and later secondary hyphae are formed. Defence mechanisms initiated following penetration include the formation of papillae (cell wall depositions) immediately below sites of hyphal advance. These papillae may retard fungal penetration or may be a prelude to the eventual cell death and necrosis which characterises the hypersensitive reaction (HR). HR occurs as a consequence of the action of the race-specific resistance genes and is considered either to act as a direct constraint to further development of the infection, or to represent the outcome of other, primary defensive responses. In barley containing a powdery mildew resistance gene, prevention of HR by a heat shock treatment results in fungal development which is equivalent to that occurring in a susceptible host.

Initiation of HR is assumed to follow the specific recognition process between fungal elicitor and host receptor molecules. Attempts to identify these molecules have been pursued assuming both to be constitutive products. Thus far, techniques involving the use of 2-D gels capable of resolving more than 300 polypeptides have been used in studies with wheat leaves and conidiospores without marked success. In barley, however, differences in esterase isoenzymes in non-infected tissue were found to be associated with varietal variation in mildew resistance, and a recent report suggests that the isolation of an *inducible* protein unique to a barley cultivar with a mildew resistance gene has been achieved.

4.2.2   Rust diseases (Puccinia spp.)

Three major rusts attack wheat: brown rust (*P. recondita*), yellow rust (*P. striiformis*) and stem rust (*P. graminis*). The latter is of less concern in Europe but has been extensively studied in the USA. *P. striiformis* and *P. hordei* also attack barley. Rust epidemics can cause very serious yield losses and changes in pathogen populations frequently necessitate the introduction of new varieties of wheat with suitable types of resistance. Both brown and yellow rusts survive continuously on wheat, and for yellow rust no sexual stage or alternate host is known. Hence, cultural measures designed to minimise sources of infection are often impractical in intensively farmed wheat areas.

Much of the information on wheat rusts concerning the inheritance and genetics of resistance and host plant responses to infection parallels that for powdery mildew. The study of other cereal rust diseases also provides useful pointers for research on those of wheat. An example is the production of phytoalexins by oats following infection with crown rust (*P. coronata avenae*). These substances, termed avenalumins, are active *in vitro* against *P. graminis* f.sp. *tritici*. They accumulate in response to infection only in incompatible host–pathogen combinations. Evidence for phytoalexin production in wheat is so far restricted to a

46

demonstration of antifungal activity in leaf extracts of Little Joss, a durably rust-resistant variety, following infection with *P. striiformis*.

Race-specific resistance to rusts is usually controlled by single, dominant genes. In wheat, some ten loci code for seedling plus adult resistance to *P. striiformis*; a further four are known which control resistance in adult plants only. Some of the loci show allelism; at least one is recessive. Chromosomal locations are known for several of the genes. Some 35 loci are known which code for resistance to *P. graminis* and one of these, Sr15, is considered to be the same as, or to overlap with, Lr20, a *P. recondita* resistance gene. Both Lr20 and Sr15 show temperature dependence in their action. It is possible, therefore, that the same gene product is involved in the recognition of races of both rust species. At least 35 Lr genes for 'low-reaction' to *P. recondita* have been named.

Species of *Aegilops*, *Agropyron* and diploid wheats have been used to introduce rust resistance into *T. aestivum*. The procedures involved are, as mentioned above, difficult and time-consuming and also involve much cytological work. In addition, resistance may be lost during the transfer due to effects of genetic background, or inhibitory genes may be present which prevent expression of the resistance gene in its new background. Nevertheless, some success has been achieved, for example, in the use of resistance from *Agropyron*.

Use of aneuploid lines has suggested that durable resistance to yellow rust might be located in the short-arm 5BS–7BS translocation chromosome of certain wheat varieties, while *loss* of the long arms of group 5B chromosomes also promotes resistance. Thus, gene deletion, provided it carries no substantial penalty in terms of undesirable characters, may be a means of increasing resistance.

Another circumstance where gene deletion would be beneficial concerns the Lr23 gene in the variety Thatcher. Expression of this gene appears to be inhibited by another gene in this variety and another example of the same phenomenon has recently been described.

Rust infection of the wheat crop normally proceeds via urediospores spread from an adjacent or preceding crop or its exposed stubble. Germination of these spores is known to be influenced by several external and internal factors. Germination is inhibited in dense spore populations by diffusible spore products: *P. graminis* produces methyl-*cis*-ferulate which is thought either to block dissolution of the germ pore plug or to inhibit early germ-tube growth. The possibility exists, therefore, that cultivars of wheat might be induced to produce an effective inhibitor of spore germination. Germination stimulants have been extracted from spores of *P. recondita* and *P. striiformis*.

Urediospore walls from *P. striiformis* contain predominantly a chitosan-like material which is also present, though to a much lesser extent, in the wall of the germ tube. Chitosan is extremely inhibitory to germination and growth of the urediospores, and this has led to the suggestion that the host plant might attain resistance if plant enzymes could be induced which attacked the urediospore walls and released chitosan.

The rust germ tubes pentrate leaves via the stomata. Germ tubes need to be correctly orientated to reach the stomata and are known to become aligned with the narrow axis of the leaf by responding to the presence of a lattice of wax crystals on the leaf surface. Wax-less mutants of the host might, therefore, show reduced infection. (The absence of diketone lipids characterises such mutants.) Growth towards, and recognition of, stomatal pores by the germ tube is thought to rely purely on physical properties of the leaf surface, although substances leached from guard cell walls may signal the development of the appressorium which occurs immediately above the stomatal pore.

As with mildews, race-specific resistance genes for rust appear to operate only after penetration has been achieved. It has been assumed that direct contact between fungal haustoria and host mesophyll cell membranes is required before specific resistance mechanisms are initiated. However, in at least two cases (see below) this may not be so. Thus, following *P. striiformis* infection of Little Joss (a wheat cultivar with 'durable' resistance to yellow rust), necrosis of mesophyll cells is observed during or shortly after the formation of the sub-stomatal vesicle (from which haustorial mother cells are produced). Also, vesicle degeneration occurs without contact with mesophyll cells. This has been taken as evidence for the production of diffusible substances by both fungus and host. Similarly, in oats, studies involving successive inoculation with two races of crown rust suggest that recognition events precede the formation of hyphae. If soluble gene products are present in the intercellular spaces of such infected leaves then their isolation by means such as infiltration washing may be possible, precluding the need to extract tissue. Washing fluid from barley leaves has been found to contain numerous proteins.

Cell death and necrosis (hypersensitive response; HR) is the most readily observed response to rust infection. In the above cases it occurs after inhibition of fungal growth becomes evident, but in other interactions, lignification associated with HR precedes or is coincident with inhibition of hyphal growth, and may be a decisive factor in such inhibition. Thus, when lignification in the presence of the Sr5 gene (a *P. graminis* resistance gene) is inhibited by either low temperature or infiltration of the leaves with spore suspensions, fungal growth is similar to that in the absence of the gene.

4.2.3 Leaf and glume blotch (Septoria nodorum and S. tritici)

These species are facultative parasites which cause leaf and glume blotch of wheat and other cereals. The diseases appear to have gained in importance in recent years and considerable research effort is now directed towards improving resistance to them in wheat varieties.

Septoria nodorum. Although numerous isolates of *S. nodorum* have been tested there is little evidence for physiological specialisation by the fungus, and hence no cultivar-specific races of the fungus are recognised. Even isolates from *Agropyron* have been found which are indistinguishable in pathogenicity to those from wheat. Generally, however, there is some degree of specialisation in that isolates obtained from wheat tend to be less virulent on barley and vice versa.

The absence of clearly distinguishable races of the pathogen means that any resistance to *S. nodorum* by wheat cultivars is generally displayed equally to all isolates of the fungus. (Nevertheless, some 'moderate' cultivar × isolate interactions have been found.) Coupled with this, resistance is often determined polygenically. Tolerance to *S. nodorum* in the variety Thatcher is associated with genes on 12 chromosomes. There is also evidence that the cytoplasm influences resistance to *S. nodorum* in wheat.

Resistance to *Septoria* is a quantitative character with continuous variation in symptom expression. It is often expressed as a reduced rate of epidemic development, equivalent to the 'slow rusting' or 'slow mildewing' phenomenon in wheat.

There is considerable variation within *T. aestivum* for resistance to *S. nodorum*. Resistance to *S. nodorum* is sometimes associated with resistance to *S. tritici*. *S. nodorum* resistance in hexaploid wheats may reside mainly on the A and B genomes, as the presumptive donors of these, *T. monococcum* and *Aegilops speltoides* respectively, show a high level of resistance whereas some lines of *Ae. squarrosa*, the source of the D genome, are susceptible.

Resistance sources in related species have not been extensively used in wheat breeding programmes. Nevertheless, several wheat relatives do show high levels of resistance to *S. nodorum*, especially the tetraploid species, *Triticum timopheevi* and *T. turgidum*, in addition to those already mentioned.

Infection by *S. nodorum* pycnospores is highly dependent on favourable environmental conditions, particularly high humidity. There are few studies on the interaction between fungus and host at the cellular level. In one study using four wheat cultivars, no differences were noted in germination or appressorium formation but the growth of the germ tubes prior to penetration varied with the host cultivar. In another study with 32 wheat varieties and 9 wild species of wheat, variation in infection frequency as well as in lesion size and degree of necrosis was found.

In infected barley, papillae (extracellular deposits) form between the plasmalemma and the epidermal cell wall beneath sites of attempted penetration. These appear to prevent further fungal growth. The papillae are initiated either in response to physical damage incurred during penetration of the cuticle by the fungus, or in response to diffusable metabolites released from it. The fungal hyphae stop growing within the upper cell wall layers before reaching the papillae. The papillae may cause lignin deposition in the host cell walls, protecting them from cell wall degrading enzymes released by the fungus. Papilla formation involves lignification, a common response to infection by fungi, and one which cannot be induced simply by mechanical wounding. In wheat, however, lignification is sometimes slow and does not always limit penetration by *Septoria* spp. In wheat varieties differing in resistance to *S. nodorum*, lignification occurs below unsuccessful penetration sites and also in the vicinity of advancing hyphae. It alone does not account for the differences in resistance observed. However, pre-inoculation with non-pathogenic fungi such as *Botrytis* which induce lignification, subsequently delays attack by *Septoria* spp.

Necrotic 'burning' in the absence of lesions on wheat infected with *S. nodorum* initially suggested production of a diffusible toxin by the fungus. A toxin which elicited symptoms of necrosis similar to those of the terminal stages of the disease in the field has been detected and partially purified from cultures of *S. nodorum*. However, the toxin, which has not been chemically characterised, is non-specific in its effects with respect to *Triticum* cultivars and species; thus the highly *S. nodorum*-resistant *T. timopheevi* is affected to the same extent as are susceptible *T. aestivum* cultivars.

It is not known whether phytoalexins are produced in response to *Septoria* infections. Constitutive benzoxazinones have been postulated to act as antifungal compounds giving resistance to *S. nodorum*, but the levels of these fall during plant ontogeny and thus do not readily account for resistance expressed in the field.

Septoria tritici. Wide variation in the susceptibility of both bread wheat cultivars and wild species to *S. tritici* has been found in studies in Israel. High levels of resistance have been observed in a number of species including *T. monococcum* var. *boeoticum*, *T. dicoccum* and *T. timopheevi.* Attempts have been made to use *Agropyron elongatum* as a source of resistance, which is known to be located mainly on chromosome VII of this species.

As an alternative to resistance, some wheat varieties may display tolerance to infection by *S. tritici* in being able to sustain grain yields despite extensive symptoms of leaf blotch. Some breeders claim that tolerance can be selected for relatively rapidly with some crosses. Both additive effects of a few loci, and joint action of several loci appear to condition the effect depending on the nature of the cross.

There appear to have been few studies of the process of infection by *S. tritici*, although again, lignification has been implicated as likely to be important in determining resistance.

4.2.4   Eyespot (Pseudocercosporella herpotrichoides)

This disease is most serious on winter wheat and barley, causing death of tillers or plants. Leaf sheaths are infected and the fungus progresses inwardly causing eventual collapse of the straw near the base of the plant. Infection is from preceding crops and their exposed straw or from infected 'volunteer' plants.

No races of eyespot are known which show marked differential virulence on cultivars of *T. aestivum*. However, races adapted to either wheat or rye are recognised. European varieties of wheat show the greatest resistance to the disease and the variety Cappelle Desprez has shown durable resistance for some 30 years. However, this may not always be adequate under extreme disease pressure. Resistance is inherited as a dominant character but genes for resistance are located on at least four chromosomes of hexaploid wheat (1A, 7A, 2B, 5D), although predominantly on chromosome 7A, and has been successfully transferred. The 7D chromosome of *Aegilops ventricosa* is an important source of resistance which has been successfully transferred into *T. aestivum*. Rye could also be a useful source of resistance, at least to wheat-adapted isolates.

Resistance of plants to eyespot has been explained by the ability of leaf sheaths and stems to resist penetration. This 'structural' resistance is, however, seen solely to be a means of limiting the rate of fungus development and does not prevent its spread. It may, however, permit the crop to mature before serious damage is done.

4.2.5   Take-all (Gaeumannomyces graminis)

Generally, infection takes place via the roots from mycelium surviving on root and stubble remains of previous crops. Susceptible weed grasses including *Agrostis* and *Agropyron* may perpetuate the disease.

Isolates have been found which preferentially attack wheat or rye. No very resistant varieties of wheat are available although Maris Huntsman shows some resistance. Dwarf wheat varieties appeared to be more susceptible than tall varieties in early studies, but work using lines near-isogenic for the presence/absence of the Rht dwarfing genes has shown that these genes do not affect resistance. Triticale is more resistant than wheat. The introduction of resistance into wheat from rye is likely to be difficult because it is probably multigenic. Use of wild relatives of wheat as sources of resistance appears to have been little explored.

Several possibilities exist for the biological control of take-all. Virus-like particles have been reported in isolates showing low pathogenicity and these appear to be absent from those displaying normal virulence. Microorganisms which antagonise take-all have been described, and they may offer a further means of suppressing the disease. Methods designed to increase the abundance or activity of such organisms, including genetic manipulation, could provide means of reducing the problem of take-all in intensively managed cereal crops.

4.3   Pests

Very many species of insects can cause damage to wheat in occasional or local outbreaks, but most live in the crop without causing significant damage under normal circumstances. The species listed below include those that most often cause significant losses in the EEC.

Aphids (*Aphididae*)
Sitobion avenae	Grain aphid
Rhopalosiphum padi	Bird cherry-oat aphid
Metopolophium dirhodum	Rose-grain aphid

Shoot flies (*Diptera*)
Delia coarctata	Wheat bulb fly
Phorbia securis	Late wheat shoot fly
Opomyza florum	Yellow cereal fly

Oscinella frit	Frit fly
Meromyza saltatrix	
Chlorops pumilionis	Gout fly
Mayetiola destructor	Hessian fly

Leaf beetles (*Coleoptera*)
Oulema melanopa   Cereal leaf beetle

Soil pests
Agriolimax reticulatus (Mollusca)   Field slug
Agriotes spp. (Coleoptera)   Wireworms

Bugs (*Heteroptera*)
Eurygaster integriceps   Sunn pest
Aelia spp.   Cereal bugs

Nematodes (Nematoda)
Heterodera avenae   Cereal cyst nematode

4.3.1 Aphids

These are important both because large infestations causing severe damage can build up during a short period of favourable conditions, and because they are vectors for virus diseases, notably barley yellow dwarf virus (BYDV). In summer, aphid populations can increase at rates that are exceptional among herbivores and the net growth of individuals during development and subsequent reproduction is, therefore, very great. This rapid growth is very dependent on the food supply and frequently is limited by changes in the plant.

Winged migrants develop when populations are numerous or on deteriorating plants, whilst wingless individuals with greater fecundity are produced on favourable plants. This combination enables aphids to invade crops very regularly and leads to large infestations under favourable conditions. Host finding by winged aphids involves vision, with orientation to yellow or green objects; scent, for example, *Sitobion avenae* detects plant odour components including 6–7 carbon atom alcohols, *trans*-2-hexenal and hexanal, and geraniol and farnesyl acetate; and taste, which is tested by brief exploratory probes of potential food plants.

In summer, the grain aphid, *S. avenae*, is more important than *Metopolophium dirhodum* on wheat, but *Rhopalosiphum padi* lives mainly on grasses. Damage is caused by the direct effects of aphid feeding and by the effects of honeydew on the leaves of the crop. Losses of yield are related to the number of aphids and the duration of infestation.

Winged *S. avenae* and *R. padi* also colonise early-sown wheat in autumn. The autumn migration into young wheat crops is most important for the spread of BYDV, of which the strains transmitted by *R. padi* are the most damaging. All three aphid species multiply on cereal crops or grasses throughout the winter when conditions are favourable. If this occurs viruses can spread widely in the crop when the plants are most susceptible.

At present, control of wheat aphids depends on insecticides, but the recent trend of applying insecticides in the autumn to prevent the spread of BYDV, as well as in the summer, significantly increases the risk that insecticide-insensitive forms will appear. No varieties of wheat on EEC national recommended lists are claimed to be aphid-resistant. However, as natural enemies often exert substantial biological control, partial resistance has the potential to give good control of aphids through positive interactions with the effects of predators.

4.3.2 Shoot flies

These have one, two or three generations per year. The biology of the different species varies considerably, but the damage they do is similar. The larvae feed on young tissue within the leaf sheaths, each larva killing one or more shoots. The effects are often not evident until after the damage has been done. Moderate levels of shoot loss, especially early in the growth of the crop, are of little significance as the plants compensate by developing more tillers. Some compensation can also occur to replace plants that are killed at an early stage, but when many plants are killed or stunted, or shoots are attacked at a late stage, the number of ears and hence the yield can be seriously reduced.

The wheat bulb fly, *Delia coarctata*, is the most important species of shoot fly in Britain. The larvae hatch in February from eggs laid in July on bare soil, and as they must feed from more than one shoot to complete development, they move in March and April to other shoots on the same or adjacent plants. The severity of damage depends on the number of larvae and the ability of the crop to compensate for the lost tillers. The movement of migrating larvae is arrested in response to root exudates from young wheat plants. Extracts from oats, which are not attacked by *Delia*, inhibit this response and, if watered onto infested wheat, reduce the number of larvae invading additional shoots.

Other shoot flies lay their eggs on or near plants, and the trend to early sowing of winter wheat has favoured those species that lay their eggs in the autumn. These include *Opomyza florum*, which has become more important in recent years in north and central Europe, and also the frit fly, *Oscinella frit*, and the gout fly, *Chlorops pumilionis*, which overwinter as larvae following autumn egg laying. As *O. frit* cannot develop on advanced stems, infestation is most severe on early autumn and late spring sowings which are exposed when young to laying adults. Gout fly is less important on wheat than on barley. It causes a conspicuous swollen gall on young tillers but will also live in the upper nodes of stems in spring.

Phorbia securis can be important in central and southern Europe. The flies overwinter as adults and larvae develop in wheat shoots in May. In southern Europe, two generations may develop, the larvae of the first early in spring in the bases of tillers. The larvae of the second feed in higher nodes, which causes fertile tillers to abort. *Meromyza saltatrix* is important in central Europe, where it causes similar damage especially as the second-generation larvae feed at the upper stem nodes.

The hessian fly, *Mayetiola destructor*, has attracted attention in the USA where it is an introduced pest from Europe. It also has autumn and spring generations, and is most damaging to young plants which are stunted by the secretions of the larvae. The delicate adult midges are short-lived and delayed sowing provides a cultural control. Varieties bred for resistance to hessian fly have given effective control of this pest in the USA, despite its ability to produce biotypes that can overcome some of the resistance genes used.

4.3.3   Leaf beetles

These include several unrelated species, but the most important is probably *Oulema melonopa*, the cereal leaf beetle. This insect has only one generation a year, but both adults and larvae feed on leaves, the latter producing characteristic pale stripes where they have eaten away the green tissue. *O. melanopa* is not important in Britain but can cause problems in central Europe.

4.3.4   Bugs

Cereal bugs are most important in south-eastern Europe and Asia Minor, where the sunn pest, *Eurygaster integriceps*, is a major hazard to wheat growing. The bugs of this species enter the crops in May and the young feed from the ears, piercing the developing grain and injecting proteolytic enzymes into it. As little as 5 per cent of such damaged grains in a sample render the whole unfit for bread-making. In southern Europe, *Aelia* species are the most important of this group of cereal bugs but are not as significant as the sunn pest. Wheat varieties with relatively large starch grains in the endosperm are less susceptible to *E. integriceps*.

4.3.5   Nematodes

Heterodera avenae is a complex species group which parasitises small-grained cereals, some grasses and maize. Coevolution with hosts has resulted in the formation of subspecies and races with limited host ranges. The form which affects wheat is a less serious pest than it was twenty or more years ago, and outbreaks occur mainly on light soils. The decline of the importance of the nematode in western Europe has been associated with the increased intensification of wheat growing and is likely to have been caused by the build-up of natural predators, including fungal pathogens. However, where wheat is subject to severe water stress or grown in extended rotations, this and related species are of economic importance.

4.3.6   Resistance to pests

In N. America, resistance has provided or contributed to control of the hessian fly, the stem sawfly, the chinch bug and the cereal leaf beetle, and resistance to greenbug is under development. However, relatively little is known about the causes of resistance at the molecular level. Resistance to stem sawfly, due to solid pith in wheat stems, and to *O. melanopa*, through epidermal hairs, exemplify mechanical constraints to pest attack that are controlled by several genes. Evidence of resistance to many pests of wheat such as the different shoot fly species is limited to observation of varietal differences in natural infestation. Resistance to wheat bulb fly damage has been associated with high tillering.

Wheat varieties with resistance to *O. melanopa* have a dense covering of hairs on the leaves. There must be at least 20 hairs each 0.20 mm or more in length per mm^2 to give effective protection. The hairs on the leaves prevent or reduce larval feeding, hairy leaves are rejected by egg-laying adults and fewer of the eggs laid on hairy leaves survive than those laid on smooth leaves. The length and density of the hairs are inherited in a quantitative manner, and the expression of the genes is affected by environmental factors. Resistance mechanisms such as this are easy for plant breeders to handle as they can be assessed directly.

Several sources of resistance to hessian fly have been exploited by conventional breeding methods in the USA. Specific interactions between resistance and fly biotypes occur with some of the genes used, but with others, combinations give resistance that appears universal. There is now evidence that about 20 genes are involved in determining resistance to hessian fly. Some single genes confer virtual immunity but for others, combinations of genes are necessary. The genes also vary in that some act only over a limited range of temperatures whilst some are temperature insensitive. Some resistant germplasms contain a number of genes. The specific relationships between hessian fly larvae and differing resistant wheat varieties are as yet little understood. It is not possible to rear or feed the larvae artificially, and hence bioassay of active compounds extracted from wheat is not possible. Larval enzymes, currently being studied in the USA, may reveal reasons for the specificity of resistance.

The resistance of wheat to aphids, in particular to the greenbug, *Schizaphis graminum*, has been ascribed to various constituents of the plant, for example hydroxamic acids. Whilst the association of these compounds with resistance is clear, it is an over-simplification to consider them solely responsible. Several different kinds of moderate resistance have been implicated in studies with *S. avenae*, including resistance associated with non-glaucousness due to lack of diketones in the cuticular waxes. Variation, including high levels of resistance, also occurs in diploid *Triticum* and *Aegilops* species. Resistance of wheat to the greenbug is being developed in the USA where this aphid was introduced from Europe. It is not a significant pest in Western Europe, although it is important in the Balkans. Different biotypes of greenbug are recognised in the USA, varying in their ability to overcome some of the known resistance sources.

4.4 Grain quality

Wheat is used for several purposes, each with different quality standards. Within the EEC, the traditional staple food of the diet, bread, is produced in many forms, by different processes, and flour suitable for bread-making in one country may be unacceptable in another. In addition, bread-making technology is not static, and process changes enable acceptable bread to be produced from flours previously considered to be unsuitable. Thus plant breeders must know what is desired for each end use and how the component properties of the grain can be modified by breeding to best meet these requirements.

About a third of the wheat produced in the EEC is used in the manufacture of animal feeding-stuffs. For this use, composition is not critical and variation among varieties is of minor importance. Cereal grains used for animal feeding have to be supplemented with protein meal to sustain the growth rates expected in intensive rearing systems. Thus, although wheat contains 10–12 per cent protein, it is regarded by compounders mainly as a source of metabolisable energy, that is starch. The amino acid composition is not important for ruminants, for they obtain their amino acids from microorganisms living in the alimentary system. For monogastric animals (poultry and pigs), where interconversion of amino acids in the gut does not take place, lysine is the first limiting amino acid in the cereal component of the feed, and attempts have been made to breed forms with a higher lysine content, particularly of barley. Despite much effort with barley over many years and some effort with wheat, no high-lysine varieties have yet been produced which have gained acceptance in agriculture. All high-lysine mutants have shrivelled grains. Selection for normal grain size, and hence acceptable yield, has always resulted in loss of the high-lysine character. No effort has been devoted to breeding high-lysine wheat. The relatively low cost of lysine produced by industrial fermentation has made it doubtful whether the effort and cost of breeding high-lysine varieties is worthwhile.

Except for a small proportion that is whole-milled, wheat milled for bread-making is required to be hard-milling and to give a high yield of white flour. Breeders have been able to select varieties with these characteristics which are strongly and simply inherited. Wheat for biscuits and cake-making needs to be soft-milling to minimise starch damage and give low water absorption to the flour, and these features are unavoidably associated with a somewhat lower flour yield. The protein content and type are especially important for bread flours. Protein content is strongly negatively related to grain yield and it has proved impossible so far to produce varieties with very high yield and protein content. The availability of rapid milling machines for small samples, and of infra-red reflectance analysers for protein content make it possible for breeders to screen large numbers of samples for protein content and will enable appropriate genotypes (and ultimately genes) to be identified if they exist. Apart from the plant breeding solution, late applications of fertiliser nitrogen can significantly increase protein concentration without reducing grain yield.

There are two main classes of protein in wheat endosperm, the gliadins and the glutenins. Gliadins are mainly responsible for the viscosity and extensibility of dough, allowing it to rise during fermentation. The gliadin in a wheat variety consists of numerous individual proteins coded for at six complex loci. Glutenin confers visco-elasticity to the dough, the elastic component preventing it from becoming over-extended and from collapsing either during fermentation or during baking. In European

58

wheat varieties, quality for bread-making is limited by poor visco-elasticity. Glutenin is a complex protein made up from component subunits. These subunits vary in molecular weight from 20 to 150 kD and are coded for by genes at six unlinked loci. Three of these are complex and also contain genes for the gliadin proteins.

The glutenin subunits differ from each other in their effect on dough strength and their allelic variation, which has been described only in recent years, can be exploited to assemble appropriate combinations depending on the dough strength required. In general, the subunits of highest molecular weight confer greatest strength to a dough. This knowledge is of great value in breeding for bread-making quality, and it is expected that varieties with greatly improved bread-making quality will be produced. Among the diploid progenitors of wheat, there are many glutenin subunits which are not present in bread wheat. Their importance for quality is not known. The allelic variants can be transferred to hexaploid wheat and, by backcrossing, breeding lines can be produced which differ only in respect of the subunit of interest. This procedure is being undertaken for a number of subunits which, on the basis of their molecular weight and structure, it is thought would be beneficial.

The enzyme α-amylase is present to varying extents in varieties, and following a wet harvest its levels can be high in varieties susceptible to sprouting. High α-amylase results in starch breakdown during dough development and in the early stages of baking, giving 'sticky' loaves with an irregular crumb structure. Selection for resistance to sprouting is effective in reducing the risk of high α-amylase, but some varieties have high levels of the enzyme in the absence of sprouting. This feature can be eliminated fairly readily by selection.

For biscuit making and most other domestic uses, a weak and extensible dough is required. In the past it has been easy to breed varieties which meet this requirement and no difficulties are envisaged in the future.

For other uses in the food industry, given a large enough demand, the protein composition could be varied within wide limits, by breeding for the appropriate proteins. It is not clear whether there is a demand for a changed amylose:amylopectin ratio (about 1:3 in most varieties), should this be required in wheat flour.

A further aspect of wheat quality is of concern to that fraction of the population at risk from coeliac disease. Susceptible individuals are allergic to some of the proteins in flour. The papillae of the upper ileum become inflamed and unable to function normally, and eventually become detached from the ileum. Individuals differ greatly in their susceptibility, but those very susceptible have to avoid all foods containing wheat and the related species barley and rye, which contain similar allergenic proteins. The molecular basis of the allergic reaction has not been established and it is not known at present whether it would be possible to produce wheat varieties lacking the particular proteins, or having them in modified, non-allergenic forms.

The role for bread-making quality of minor flour constituents, such as lipids and proteins which have functions other than storage, is at present imperfectly understood. These constituents are now being studied and it may be found desirable to change their amounts or composition.

4.5   Straw quality

As noted at the beginning of this chapter, numerous uses for wheat straw have been proposed. For many of these, the composition of straw as produced at present may be quite adequate. Little is known about genetic variation in straw composition and so the scope for making changes by conventional breeding cannot be assessed. Possibly the greatest scope for increasing the value of straw is to modify it to increase its value as an ingredient of feeding-stuffs for ruminants. This is a complex issue because straw is very heterogeneous and the different anatomical fractions vary in their digestibility. Hence there is scope for increasing digestibility by modifying the ratios of leaf:sheath:stem:chaff either by breeding or by mechanically fractionating the harvested straw. In making changes in straw composition by breeding it would be necessary to check that there were no detrimental effects on grain yield, for example as a consequence of a greater tendency to lodge if selection for reduced lignin content were sought.

4.6   Molecular biology and the improvement of wheat

4.6.1   Improvement of yield

A number of major genes affecting development and height have been identified and all are important for best adaptation to particular environments. However, they must represent only a very small proportion of the 'useful' genetic variants, many of which have been accumulated in varieties as a result of selection probably for several thousand generations. Although it could be valuable at some stages in breeding programmes to be able to recognise the major genes referred to above, appropriate alleles can fairly readily be fixed in breeding material, and most European programmes probably have the required allelic variants.

60

Despite this, knowledge of the number of loci and alleles regulating sensitivity to photoperiod and vernalising temperatures is very incomplete, partly because it is difficult to recognise specific alleles. Also, the molecular basis of response to photoperiod and vernalising temperatures is not understood for wheat or other species, and is a major unresolved aspect of plant biology. It may be concluded, therefore, that only when this basic knowledge is gained, will it be possible to develop rapid methods for characterising allelic variation. In turn, only when allelic variants can be recognised will it be possible to measure their effects on development and specify which alleles a variety should possess for best performance in a given environment.

Genes which affect growth rate, through their effects on either temperature responses and respiration rates or photosynthetic properties, are less well understood than are those which affect development. There is little doubt that wheat varieties from different regions of the world display variation in their growth rate response to temperature, and that spring wheat varieties grow more rapidly than do winter ones. Though there appear to be exceptions, it is widely believed that faster growth rates in winter time, associated with low soluble carbohydrate status in the tissues, render such plants more susceptible to low temperature injury. It is not clear whether genotypes with below average growth rates at low temperatures in winter also have low growth rates at higher temperatures and at later stages of their development or whether growth responses to temperature at different stages of growth are under independent genetic control. In practice it is difficult to make such assessments because growth rates are much influenced by developmental patterns, which are determined by other genes and by the prevailing temperature and photoperiod. At present, therefore, we do not have sufficient knowledge of the physiology and genetics of growth and development to identify 'target' genes for modification and/or transfer.

Respiration rates are closely linked to growth rates and, in growing tissues, to the levels of soluble carbohydrates (mainly sucrose). It seems likely that where environmental factors favour growth (i.e. the increase in plant structural material), respiration proceeds at maximum efficiency, whereas when the environment is less favourable, and in senescent tissue, respiration is less efficient. Respiratory carbon metabolism occurs by two pathways, glycolyis, and the reductive pentose phosphate (RPP) pathway. As far as is known glycolyis predominates (70% of the respiratory carbon flux), but it is uncertain whether the fluxes through the two pathways can vary with genotype, environment or stage of growth. This is partly due to the great difficulty of making reliable measurements.

61

Once the carbon has reached the Krebs (tricarboxylic acid) cycle, reducing equivalents are transferred from it to the cytochrome chain or to an 'alternative', cyanide-resistant pathway, considered to be less efficient in terms of the generation of reducing power and known to be non-phosphorylating. Except in very specialised tissues (not present in most crop plants) the cyanide-resistant pathway appears to function as an energy overflow system when levels of soluble carbohydrate are in excess of those required for growth. Attempts to eliminate the cyanide- resistant pathway, therefore, may be unrewarding. On the other hand, selection for more rapid growth (if desirable for other reasons) would result in the pathway being little used, and in more efficient respiration. In *Lolium*, selection for low rates of dark respiration has resulted in increased dry matter yields, though it is not known whether the RPP pathway has been affected. Thus, more evidence is needed before it can be concluded that attempts to modify the RPP pathway by molecular biological methods would be beneficial.

With the proviso that the demand for carbon for growth is not less than the assimilatory capacity of the plant, it will generally be beneficial to increase the assimilatory capacity. As noted earlier, assimilatory capacity is a function of the photosynthetic properties of unit leaf area, of the total leaf area and of the distribution of the intercepted light over this leaf area. The proportion of the plants' resources invested in leaf area (strictly photosynthetic 'machinery') varies greatly with the stage of growth and depends to some extent on the environment. However, for plants at comparable stages of growth and in a common environment there appears to be relatively little variation in this proportion. Hence, genetic differences in assimilatory capacity will depend mainly on genetic differences in efficiency of the photosynthetic machinery.

Where attempts have been made to ascertain whether yield improvements made by breeding can be related to improvements in the photosynthetic machinery, it is often found that photosynthetic rate per unit leaf area is similar, or less, for modern, high-yielding varieties than for their predecessors. This is true for cultivated bread wheat when compared with its wild diploid ancestors. Bread wheat has larger, longer-lived leaves than its ancestors, but a light-saturated rate of photosynthesis some 25–30 per cent lower. Genetic analysis of this difference is complicated because of differences in ploidy and cytoplasmic as well as possible nuclear differences. At present, the experimental evidence suggests that nuclear genes affecting photosynthetic properties from the A genome of the wild ancestor *T. urartu* are different from the corresponding genes in bread wheat. If this can be confirmed, research should be directed to examining variation in those proteins of the photosynthetic machinery known to be nuclear-encoded. Such studies might provide a biochemical basis for screening hybrid populations derived from crosses between the diploid *T. urartu* and bread wheat and among diploid wheat species varying in their photosynthetic properties. It must be pointed out, however, that the differences in photosynthesis may be caused not by variation in those proteins which are constituents of the light harvesting apparatus and/or of the stromal enzymes involved in photosynthetic carbon metabolism, but by variation in those proteins which have a regulatory role. There is very little knowledge of proteins (or other molecules) which exert such regulation. They are likely, however, to be minor constituents of chloroplast protein, and it will be difficult to identify which of the many candidate proteins are significant in determining the genetic variation.

Of the major proteins, the possibility of modifying the relative rates of carboxylation and oxygenation of ribulose bisphosphate by the enzyme ribulose bisphosphate carboxylase/oxygenase (Rubisco), remains an enticing target for biomolecular engineering. An increase in the rate at which it carboxylates the substrate, relative to its oxygenase activity (carboxylase/oxygenase) might increase net photosynthesis. The actual ratio of carboxylase/oxygenase depends on the relative partial pressures of carbon dioxide and oxygen in the stroma of the chloroplasts where the enzyme is located. Increasing the carbon dioxide concentration or decreasing the oxygen concentration increases the carboxylase/oxygenase, increases photosynthesis and the growth rates of C3 species (which include wheat, oilseed rape and faba beans). However, in normal atmospheres the oxygenase activity, feeding the photorespiratory glycolate pathway, protects the chloroplasts from damagingly high oxygen concentrations. Therefore, if it were possible to modify the enzyme to increase its carboxylase/oxygenase ratio, such change could adversely affect the plant through high oxygen concentration. To counter this effect, changes in the metabolic pathways consuming oxygen would be needed.

Much effort is now being devoted to studying the Rubisco protein. It consists of eight large subunits which are coded for by the chloroplast genome and eight small subunits which are nuclear encoded. There is one active site on each large subunit and both the carboxylase and oxygenase reactions cleave ribulose bisphosphate at the same point, between C-2 and C-3. The function of the small subunits is not known, though, being nuclear encoded, they are presumably potentially more variable in structure and possibly function than the large subunits. The enzyme requires carbon dioxide and magnesium ions to activate it, but its activity can be modified by several other molecules present in the stroma of the chloroplast. The molecular basis which determines the carboxylase/oxygenase ratio is not yet understood. Much more effort to elucidate the structure and function of this enzyme is needed. The results of this research will show whether it will be possible to modify the enzyme to increase photosynthesis at the leaf and crop level.

It has been noted earlier that shortage of water frequently limits wheat yields. The exploitation of plant characteristics that increase the ratio of carbon assimilated to water transpired, provided that they did not have other disadvantageous effects, would enable increased dry matter yields to be obtained for a given limited amount of available water. The need is not to maximise the ratio, as this might result in an unacceptably low rate of photosynthesis and growth, but to optimise it to give the maximum crop photosynthesis for the amount of water available.

The ratio of carbon assimilated to water transpired (the assimilation ratio) depends on the environment, notably the atmospheric vapour pressure deficit and the carbon dioxide concentration, and varies with genus. In a given environment, the ratio depends on the photosynthetic capacity of the leaves and on the stomatal conductance. Ways in which photosynthetic capacity may be increased have been discussed earlier in this section, and it has been noted that high capacity would increase the assimilation ratio. Stomatal conductance is regulated by the plant in response to the prevailing environment. Although it has not been established unequivocally, it appears that there may be scope for increasing the assimilation ratio in wheat by modification of the processes which determine the stomatal responses to the environment. Because transpiration is reduced proportionately more by a given degree of stomatal closure than is photosynthesis, the assimilation ratio is increased more than the rate of photosynthesis.

The 'signal' to which the plant responds by closing its stomata is probably a reduction in the turgor pressure of the cells near the stomata, and this signal is generated when the rate of transpiration exceeds the capacity of the plant to replace the transpired water. In turn, loss of turgor increases the rate of synthesis of the hormone abscisic acid, which, through its effects on the ion flux into the guard cells, leads to stomatal closure. Abscisic acid, formed in stressed leaves and other green tissues, is also translocated in the phloem. On reaching plant parts where cell expansion is occurring, abscisic acid reduces the rates of expansion and the final sizes of cells, and hence organs. It is uncertain whether there is a direct effect of abscisic acid on the photosynthetic apparatus other than via the stomata or possible feedback mechanisms brought into

64

operation as a result of the reduced growth of developing organs, and hence lower demand for the products of assimilation.

It seems likely that genetic variation, both in the production of abscisic acid in response to water stress and in the sensitivity of the various processes to the concentration of the hormone, will affect the assimilation ratio and growth when water is limited. Capacity to synthesise the hormone varies among wheat varieties and is a strongly inherited character. Selection experiments have shown that a 1.5- to 2-fold increase in the capacity to synthesise the hormone increases the assimilation ratio of the crop. In relation to this aspect of drought resistance, more needs to be learnt about the processes involved, and their consequences for water use and yield in a range of environments, before opportunities for biomolecular engineering can be discerned. Molecular biology can, however, aid in the identification of the genes and their products which are responsible for the synthesis and metabolism of abscisic acid, and those which are involved in determining responsiveness to it.

Another adaptive response to water stress is the accumulation of solutes in cells which helps to maintain turgor as cell water potential decreases. This phenomenon, osmoregulation, is probably more important in some xerophytic species than in most crop plants, including wheat, which are mesophytes. In crop plants, osmoregulation may serve to maintain photosynthesis and growth during relatively brief periods of water stress, in the range of hours or a few days. However, there is likely to be a metabolic 'cost' of this protection, because it involves the synthesis and subsequent metabolism of osmoregulatory molecules. Therefore, it is likely that there will be an optimum osmoregulatory capacity. In wheat, the major solutes which accumulate during stress are sugars, amino acids (especially proline) and glycine betaine. Organic acids and inorganic ions do not appear to be accumulated in substantial amounts in wheat, in contrast to some other plants. In unicellular algae, sugar alcohols and glycerol are accumulated. In halophytes, betaines and proline are important. To be effective in osmoregulation, solutes must be very soluble – hence a low molecular weight is favoured; they must carry no net change at neutral pH; they must be retained by the cell's plasma membrane against a large concentration gradient; they should cause minimum alteration to the structure of water and to the extent that they do, they should stabilise enzyme structures in their active forms; and their concentration must be under fine control by a turgor-sensing mechanism.

In addition to molecules which are actively synthesised in response to drought (as distinct from those which are merely concentrated as a consequence of water loss), it may be envisaged that other molecules, present at normal water potentials, can provide some protection against the effects of water loss. Molecules of this kind function as cryoprotectants in some specialised plants and animals, but it is not known whether molecules with a similar function occur in xerophytes.

At present there is no clearly definable target for biomolecular engineering which would or might lead to an increase in the drought tolerance of wheat. More information is needed on genetic variation in the capacity to accumulate the various osmoregulatory molecules, among wheat and related species, to ascertain the biochemical basis of this variation and its genetic control. In addition, if molecules which confer resistance to stress can be identified in xerophytic species, and the genes coding for their synthesis identified, these genes would be candidates for transfer to wheat.

Finally, in relation to increased drought tolerance, it should be noted that although it would be beneficial in the EC countries, it would be of much greater value in most other wheat-growing countries of the world, where drought is a much more severe limitation to yield.

In summary, among many possible applications for molecular biology, the following seem to be particularly pertinent to wheat:

1. Vernalisation and photoperiod responses as model systems of environmentally regulated gene expression.
2. Studies of the molecular basis of differences between *Triticum* species in their photosynthetic characteristics, with particular attention to light harvesting chlorophyll proteins and to the small subunit of Rubisco.
3. Genes involved with and activated by abscisic acid and by water stress, and their protein products.

4.6.2   Improvement of resistance to disease

Because they are deemed to be convenient model systems, pathogens and hosts which are of only minor economic importance are often chosen for studies on the molecular basis of resistance. The diversity in both pathogens and hosts is such that much of the information gained may be of little direct relevance to pathogens of major importance. Accordingly, there is much to be gained by focussing the attention of pathologists and molecular biologists on such pathogens. For wheat in the EEC, these are yellow and brown rust, powdery mildew, take-all, Septoria and eyespot.

For the rusts and mildews, where there is much diversity in races, a rapid means of identifying races is needed. Such a system could be based on restriction fragment length polymorphism of the fungus DNA or an antibody system, or possibly on cDNA probes.

To gain a better understanding of resistance mechanisms attention needs to be paid to the initial recognition (or non-recognition) phenomena. One possibility would be to search for molecules in the host which interact with those in the pathogen. Such molecules are likely to occur at cell surfaces and to be polymers. Where there is a considerable diversity in pathogen virulence and host resistance, it may be supposed that there will be a corresponding variability in the recognition molecules, which could be exploited to improve resistance. Once such molecules have been found, similar molecules in related resistant species or genera could be sought with a view to comparing their structure and transferring the genes coding for them to the crop plant of interest. Parallel approaches could be made by developing cDNAs to sequences in higher plants that code for recognition molecules.

Another approach is to study the regulation of phytoalexin production so that the genes involved could be identified and variability sought. The objective would be to enhance phytoalexin production in response to infection, thus preventing the spread of pathogens.

4.6.3   Improvement of resistance to pests

Molecular biological techniques could be used to introduce a general resistance to the shoot fly pests in wheat. These flies present chronic and apparently intractable problems. Conventional methods would involve identification of resistance to each species, with limited prospects that resistance to one would affect others. Considered separately, these pests hardly warrant the effort involved. Relatively little is known of their feeding or, with the exception of the hessian fly, of resistance to their attacks. In contrast, development of a general resistance to all species would give a character of substantial value throughout the EEC.

To do this, attention should first be given to the possibilities of exploiting the Diptera-specific toxins from *Bacillus thuringiensis* and closely related species, for example the serotype H14 toxin from *B. israelensis*. It would be necessary to identify a toxin type with a suitable range of activity, but if one can be found, the background work with *B. thuringiensis* and the toxins it produces should give a basis for rapid progress.

The positive response of bulb fly larvae to wheat extracts and the negative effects of oat extracts indicate that these insect–plant associations might easily be broken if molecular methods could be used to introduce alien variation to the wheat plants. It seems likely that oats and wheat are sufficiently similar for the oat genes to function in wheat. It is most probable that other substances could deter or destroy invading larvae of bulb fly and other shoot fly species.

A second possibility, therefore, is to consider the introduction of genes to generate alien secondary substances in a plant. Transfer from oats to wheat would affect bulb fly, and probably gout fly, *Opomyza florum* and *Meromyza saltatrix*, but would be most unlikely to reduce attacks by frit fly or *Phorbia securis*, as these live on oats naturally. Introduction of substances that do not occur naturally in Gramineae might well yield wheat with resistance to all these pests.

Little is known of the sources of resistance to *Sitobion avenae* that have been identified, but in general these do not protect against other aphids. It is likely that the precision in the feeding behaviour of aphids may allow them to attack some transformed plants (e.g. with *B. thuringiensis* genes), in the same way that the peach-potato aphid, *Myzus persicae*, can feed from tobacco even though it is susceptible to nicotine.

Should wheat cyst nematode become a more significant pest, a search for resistance genes in *Triticum* germplasm would be desirable. Associated with this, it would be important to have a rapid and precise method for identifying any virulent pathotype or member of the *Heterodera avenae* species complex. Since there may be no anatomical or histological differences among pathotypes, a number of biochemical tests have been suggested. The most direct way of identifying differences between pathotypes is by the mobility of the proteins extracted from them as visualised by electrophoretic procedures. A development of this is to prepare antibodies to one of the proteins shown to differ between two pathogens and to use immunological methods (RIA or ELISA) to classify the nematodes. The use of RFLPs (Section 3.2) does not seem feasible at present because of the large amount of nematode material that would be required for each test.

4.6.4   Improvement of grain quality

Recent developments in the understanding of the genetics of the endosperm proteins have made it possible to produce high-yielding varieties which are much more suitable for bread-making than was previously considered possible. In principle, it would be possible for breeders to handle larger populations if quality characters could be selected for at the early vegetative stage instead of in the seed. This is not possible at present because the proteins, rather than the genes, are detected, and the proteins are only present in the endosperm. If cDNA probes specific for allelic variants of the relevant genes could be developed, these could be used for testing individuals in segregating progenies, so that only those containing the desired alleles need be retained. For all other characters attention could then be directed to the retained individuals. However, several tests, each using a different probe, would be needed, whereas electrophoretic separation of the proteins in the endosperm of half-grains is being used to detect in one test all the proteins of interest (the half containing the embryo can be grown on to produce a plant). Therefore it is doubtful whether tests based on such cDNA probes would be useful in breeding programmes. It would probably be easier to prepare highly specific monoclonal antibodies to individual storage proteins than cDNA probes for them. With the antibodies it might be easier to detect the presence or absence of particular proteins than by electrophoresis.

The molecular basis of protein quality in bread-making is being actively studied and the key proteins have been identified. As stated earlier, the various high molecular weight (HMW) glutenin subunits play a major role in producing a dough of sufficient strength for bread-making. Natural mutants lacking certain HMW subunits cause a decrease in the proportion of these key proteins as a percentage of the total endosperm protein and a dramatic decrease in quality for bread-making. Assuming gene copy number to be proportional to the output of HMW subunits, insertion of additional genes coding for suitable HMW subunits into wheat by molecular biological means should increase quality. Since the genes already occur in wheat, it would be expected that the inserted genes would come under the control of existing regulatory genes so that they are de-repressed in the correct organ of the plant and at the right time. Quality for bread-making, and particularly the keeping quality of the bread, is strongly related to protein content. The total amount of protein synthesised in the grain is limited by the supply of amino acids from the vegetative organs and so a general increase in the copy number of the genes coding for endosperm storage proteins is unlikely to result in increased protein content.

Another way in which molecular biologists could help the breeder develop new varieties with improved bread-making quality would be to advise them which protein genes to insert from species related to wheat by conventional cytogenetical procedures. Currently, the genes for HMW glutenin subunits which are associated with good bread-making quality are being sequenced and it is anticipated that the biochemical basis for the differences in quality may be elucidated. Restriction endonuclease digestion methods could be devised so that good-quality subunit types give specific electrophoresis patterns. The vast number of homologous proteins could then be analysed from the wild diploid relatives of wheat. Types suspected of having the correct biochemical structure for conferring strong elasticity to a dough could then be transferred to modern wheat by conventional cytogenetic procedures.

69

5

Oilseed Rape

Rapeseed oil provides only about 10 per cent of the world's production of vegetable oils (Table 5.1). Although major modifications in its composition achieved by breeding have removed most of the undesirable constituents, prejudice against rapeseed oil remains. The crop is well adapted to north-western Europe. The production in the EEC of other oilseed crops, notably sunflower and soya, could compete with that of rape if and when the Community aims to become self-sufficient in vegetable oils. Thus, in assessing the feasibility or desirability of increasing the production of rape seed for oil, the same assessments need to be made for sunflower and soyabean, the production of which is being

Table 5.1 *World production of oil and meal from vegetable sources in 1980 (millions of tonnes (Mt))*

Source	Oil	Meal
Soyabean	15.0	60.0
Palm and palm kernel	4.5	0.6*
Rape	4.0	6.0
Sunflower	4.0	5.0
Coconut	3.2	1.8
Cotton seed	3.2	10.0
Groundnut	3.2	4.0
Olive	2.0	—
Linseed	1.0	1.4
Sesame	1.0	1.4
Castor	0.45	—
Safflower	0.30	—
Total	41.8	89.8

* From kernels only.

increased in the EEC. However, these questions are beyond the scope of this report and are not considered further.

As with wheat, the production of oilseed rape in EC countries has been greatly stimulated by the Common Agricultural Policy. Although oilseed rape was a significant crop in north-western Europe until the middle of the nineteenth century, the development of plantation agriculture coupled with cheap transportation led to the rapid development of tropical oilseeds and a corresponding decline in the production of oilseed rape in Europe. More recently, extensive agriculture in Canada has provided Europe with oilseed rape and its products. Importation has declined greatly in recent years as European production has increased. In 1982, production in EC countries was about 2.7 Mt (seed) per year, compared with 0.3 Mt in 1961. At an average oil content of 40 per cent, this corresponds to an annual production of rapeseed oil of about 1.1 Mt. Of the vegetable oils consumed mainly by humans in the EC countries in 1981 (3.6 Mt), rape constituted some 0.6 Mt, the remainder being soya (1.5 Mt), olive (0.9 Mt) and sunflower (0.6 Mt). A further 1.1 Mt was used for purposes other than human consumption (e.g. for the manufacture of soaps and cosmetics). Although the EC countries were virtually self-sufficient in rapeseed oil in 1981, nearly all the soya oil and much of the sunflower oil was imported, together with more than 0.6 Mt of other oils (palm oil, palm kernel oil, cotton seed oil, ground nut oil, etc.), so that the EC self-sufficiency in vegetable oils was only 33 per cent. The corresponding figure for 1982 was 39 per cent. Recently, the production of sunflowers has greatly increased in France, providing the prospect of increased self-sufficiency in this high quality vegetable oil.

Thus, while vegetable oils cannot be substituted for each other for all uses, it would seem that there is scope for the import substitution of soyabean oil and some tropical oils by increasing the production of rape seed oil. In the long term, this would be commercially feasible only if rapeseed oil were equal to or lower in cost to the users than the competing oils.

Rapeseed is also valuable for its meal, which is used at low inclusion rates in compound feeds, mainly for cattle. Although the meal has a good amino acid composition for feeding to monogastric animals (pigs and poultry), little or none is used for this purpose because of toxic and other undesirable constituents in the meal. There is one good prospect that these constituents can be eliminated, and so increased production would reduce the need for the importation of protein concentrates from non-EEC sources.

71

5.1 Biological limitations to yield

In the UK, France and the Federal Republic of Germany (FRG), winter varieties are grown. Winter kill is rarely a problem in the UK, and only occasionally results in crop losses or failure in central France and in the FRG. The winter crop is sown in August and so yields benefit from the much longer growing season than is available to the spring-sown crop (sown in April and harvested in August–September in the UK). The winter crop often experiences drought after sowing, especially in continental Europe, and this reduces the establishment and growth rates of seedlings. There may be little scope for genetic improvement in features which overcome this limitation, though screening genotypes for the ability to germinate at low soil water potential should be easy. Most probably, improved cultural methods could be devised to avoid or reduce the problem. An alternative solution would be to sow later, following rains in the autumn but, to maintain the yield, forms would be needed which had the capacity to grow more rapidly from these late sowings. This might be linked with reduced winter hardiness, however, and so might prove to be an impractical solution.

Summer drought during seed development, though its effects are less well quantified than for wheat, is likely to be a limiting factor, especially in France and the FRG. Little is known about the scope for breeding for resistance to drought by selecting for particular characters. Unlike cereals, which have narrow, adventitious roots which do not normally extract water from below 1.5 m, rape has longer roots, so it should be possible to select for deeper rooting capacity, thus enabling the crop to draw on a larger reserve of soil water and suffer less from drought. In view of the difficulty of selecting directly for rooting habit, empirical selection for high yield in drought-prone climates may prove to be the most effective means of reducing this limitation.

The average yield of seed in the five main EC countries where the crop is principally grown was about 3 t ha^{-1} in 1982. The energy required for the synthesis of rape seed is about 1.6-fold greater than that required for a similar weight of wheat grain. Hence the 'wheat equivalent' of 3 t ha^{-1} of oilseed rape was 4.8 t ha^{-1}, compared with the average yield of wheat in the same five countries of north-western Europe of about 6 t ha^{-1} (in 1982). These calculations suggest that the energy yield of rape seed is about 80 per cent of that of wheat grain. However, energy yield is not a reliable means of comparing the physiological performance of the crops. For this, knowledge of the carbon economy of the crops is needed. While there is adequate knowledge of this for wheat, this is not the case for rape, and only a crude calculation can be made. The calculation gives a potential yield of seed of about 7 t ha^{-1} for the present plant type. It must be emphasised that given the assumptions made, seed yield would be increased if the period of pod growth was increased, if the petals were absent, so permitting more radiation to be absorbed by photosynthetically functional tissue, or if the efficiency of photosynthesis (carbohydrate produced per unit of photosynthetically active radiation absorbed by the leaves and green surfaces, net of all respiratory losses) were to be increased.

In common with many seed crops, rape produces many more potentially viable ovules than eventually develop into seeds. The calculated potential yield of 7 t ha^{-1} assumes that neither the numbers of seeds nor their capacity for growth impose any limitation on yield, that is that yield is supply-limited. It may be, however, that at least in some seasons, yield becomes demand-limited as a consequence of the abortion of pods, or of ovules. It is suspected that this may occur during flowering, when the petals absorb up to 60 per cent of the solar radiation. As a consequence, the supply of carbohydrate for pod and ovule growth may be insufficient and a proportion abort to maintain the balance between supply and demand. If after petal fall the photosynthetic capacity of the crop can exceed the capacity of the remaining pods and ovules to utilise carbon skeletons for growth, the crop would become demand-limited. Probably the only rigorous method for assessing whether yield is restricted in this way is to breed forms of the crop with reduced petal size which intercept less radiation at flowering. Variation for this character exists and attempts are being made to produce experimental genotypes contrasting in petal size.

Disregarding a possible loss of yield which may result from a transient shortage of carbon skeletons during the early period of pod growth, large plants at flowering time, by providing a larger capacity for supply, and a larger demand, make for high yield, up to a level of at least 5 t ha^{-1} of seed. Mutual shading will presumably impose an upper limit to yield, though, as with wheat, it may be expected that this can be modified by breeding for improved uniformity of light distribution within the canopy, for improved photosynthetic efficiency and for reduced respiratory loss. For these features the arguments presented earlier for wheat (Section 4.6.1) apply. In the belief that more upright pods would improve the distribution of light within the canopy during the period of seed growth, breeders are selecting for more erect pods. The consequences of this selection for yield have not yet been assessed, however.

Two other constraints on yield are recognised by breeders and agronomists. Lodging, which occurs during heavy rain especially when

73

accompanied by strong winds, can cause serious loss of yield. It is probably more common in England than in other EC countries. In lodged crops a proportion of the pods is close to the ground and the affected pods rot. Also, secondary vegetative growth and flowering occur, which complicate harvesting and increase losses of seed during combining. Tall varieties and high-yielding crops are most prone to lodging, and where lodging is severe the most lodging-resistant lines give the greatest yields, implying that in these circumstances lodging is a major limitation on yield. Genetic resistance to lodging is likely to assume greater importance as varieties are produced which have a higher yield potential, and are grown with higher inputs. Genetic resistance to lodging is likely to be complex, but thick stems and taproots, with well-developed lateral roots providing good 'anchorage' in the soil, together with reduced height, are likely to be important component characters.

The other constraint, shattering of pods (i.e. their dehiscence before and during the harvesting of the crop), also causes losses which vary greatly with season and are difficult to quantify. In the worst cases, shattering is considered to cause losses of up to 50 per cent. Experimentally determined losses average about 10 per cent. Losses are the result of the normal drying out of the pods which result in the walls separating from the false septum. The separation is triggered mechanically by wind action or vibration. About half the crop in the UK is cut and swathed ten days or more before combining so preventing movement induced by wind and reducing shattering. Nevertheless, shattering remains a problem. Varieties are reported to differ in the extent to which there is a bridge of sclerenchymatous tissue between the pod walls and the septum, but the genetic control of this character is not known. Recently, a plant bearing pods with four to six instead of the normal two valves has been found. In this plant, and its progeny, the valves are attached to a central plug of fibrous tissue which extends from the base of the pod to at least half its length. Such pods appear to be more resistant to shattering, but it remains to be ascertained whether the two recessive genes which determine the character have undesirable pleiotropic effects, or are closely linked to deleterious genes.

Because oilseed rape is a dicot, like many weed species, the basis of herbicide selectivity relies on smaller biochemical differences than is the case for most herbicides used on cereal crops. In Canada a mutant resistant to the herbicide atrazine was found in a wild *Brassica*. It is maternally inherited and the resistance is probably due to a single amino acid change (serine to glycine) in the 33 kD atrazine-binding protein of photosystem II. In Canada, repeated backcrossing into spring rape has

74

failed so far to give a variety which is as high-yielding as the recurrent parent. This may mean that the change in the 33 kD protein confers reduced fitness, or that there is a linked change elsewhere in the chloroplast genome which is responsible. Reduced fitness seems to be the most likely explanation for it has been shown that a mutant of *Chlamydomonas* which has a similar (serine to alanine) amino acid change in the 33 kD atrazine-binding protein has a slower rate of electron transport out of photosystem II than the wild type.

5.2   Diseases

Increased intensification of this crop in the areas where it is grown has probably aggravated disease problems. Diseases regarded as being important are: canker (*Leptosphaeria maculans*), light leaf spot (*Pyrenopeziza brassicae*), downy mildew (*Peronospora parasitica*), dark leaf spot (*Alternaria brassicae*), clubroot (*Plasmodiophora brassicae*) and stem rot (*Sclerotinia sclerotiorum*). Diseases of oilseed rape are common to other *Brassica* and *Cruciferae* species; hence much of the information on related species is relevant. *B. napus* (genome AACC) is regarded as a synthetic species derived from *B. oleracea* (genome CC) and *B. campestris* (*syn. B. rapa*) (genome AA). Gene transfer from related species is therefore often a feasible strategy for increasing disease resistance using traditional breeding methodology.

5.2.1   Stem canker (Leptosphaeria maculans)

At the present, this is the most important disease of oilseed rape in the UK although the leading varieties grown show resistance. The crop is most susceptible to the disease in the seedling stage and adult plant resistance is thus more easily obtained. Infected seed is known to be a source of disease.

Race-specific resistance against this pathogen is not well documented. However, there is pathogen variation or, alternatively, varietal interaction with environment, as oilseed rape cultivars resistant in Europe have been found to be susceptible when grown in Australia. Also, isolates from different Australian states differed in pathogenicity when compared directly. There is varietal variation in the degree of resistance among rape cultivars. The occasionally very resistant plant has been found in an otherwise susceptible cultivar.

Adult plant resistance found in the A genome of *B. juncea* has been transferred to *B. napus*. This resistance is dominant. Seedling plus adult resistance genes appear to be located in the B genome of *B. juncea* and hence cannot be transferred by crossing.

There is little information on the physiological responses to infection or on mechanisms of resistance. Increased thickening of stems involving lignification has been noted in response to infection and imparts a general, mechanical resistance to canker. Physiological and other factors involved in the increased resistance associated with ageing do not appear to have been investigated.

5.2.2 Light leaf spot (Pyrenopeziza brassicae)

This is a serious disease in low erucic acid cultivars of oilseed rape. Most varieties which are resistant have a high erucic acid content but there are exceptions. Little is known of the genetics of resistance or of pathogen variation. Infection by the conidiospores is aided by the use of herbicides, which reduce the amount of epicuticular wax on the surface of the leaves; hence cuticle composition might be examined as a basis for general resistance.

5.2.3 Dark leaf spot (Alternaria brassicae)

This disease has been only an occasional problem. No adequately resistant varieties have been produced but there is variation in susceptibility. Partial resistance is conferred by a thick, water-repellent cuticle. There is some evidence for production of host-specific toxins by the fungus, and this could provide a basis for resistance if enzymes neutralising the toxins were induced or receptors necessary for their action eliminated from the plant.

5.2.4 Clubroot (Plasmodiophora brassicae)

Success has been obtained in transferring clubroot resistance genes from *B. campestris* to *B. napus* and from *B. oleracea* to *B. napus* and vice versa. Tissue-culture techniques have been developed for oilseed rape which offer the further possibility of introducing resistance genes from *Raphanus* into *Brassica* species. At present Dutch stubble turnips (*B. campestris*) are a major source of resistance to clubroot and are being exploited in breeding forage rapes. Resistance in *B. napus* is conferred by dominant non-allelic genes, thought to be mainly located on the A genome. (It is probable that the same genes are present in *B. campestris*.) Several such resistance genes have been accumulated in adapted oilseed cultivars in Canada, but the need for general (race non-specific) 'horizontal' resistance still exists.

Pathotypes or races of *P. brassicae* are said to be easily identified using a standard set of differential varieties. However, single spore isolates have not generally been used in such studies, so that it is unlikely that precise characterisation has been achieved.

Little is known of the mechanisms of resistance to clubroot. Rapid papilla formation has been observed following initial infection by the specialised penetration apparatus (the stachel) which is ejected from an attached zoospore into and through the hair cell wall. Root hair infection and subsequent release of secondary zoospores appear to be unrelated to race-specific resistance. There remains the need to determine how soon after root infection such resistance is expressed.

The gall-like growth resulting from infection is due to a stimulation of host cell division and reflects a disturbed hormonal metabolism of the tissue. It was initially proposed that *P. brassicae* induced the formation of auxins from gluco-brassicin, a constituent common in Crucifers. However, cytokinin as well as auxin levels are high in infected root tissues compared with normal root tissue. Infected root-gall grown in tissue culture is independent of exogenous auxin and cytokinin but this property is lost when the gall is cultured in the absence of the pathogen. Hence, hormone production depends on the continued presence of the fungus.

Direct infection of stem embryonic haploid tissue cultures of oilseed rape, following which the fungus completes its life cycle within the cultured cells, offers a promising means of studying infection and development under controlled conditions and possibly selecting resistant lines. It has been suggested that screening might be based on the hormone requirements of infected tissue, if differential requirements can be shown to exist. Also, tissue cultures can be conveniently exposed to mutagenic agents and could subsequently be screened for any resistance arising from mutagenesis.

As an alternative to increasing host plant resistance, *P. brassicae* might be controlled by accelerating the degradation of cell walls of its resting spores – which otherwise remain viable for long periods in undisturbed soil. The possibility of selecting (or genetically engineering) bacteria capable of releasing suitable degradative enzymes in the vicinity of the spores might be considered. Use of various additives applied to soil, designed to stimulate such a process appears to have met with limited success.

5.2.5 *Stem rot* (Sclerotinia sclerotiorum)

This disease is of major importance in France and has occasionally been severe in the FRG. Its importance could increase if the frequency of oilseed rape and peas in arable rotations increases, for peas are affected by the same organism. No genetic source of resistance is known. However, as the normal method of infection is through ascospores germinating under petals which have fallen into leaf axils towards the end of flowering, apetalous types might be less affected.

5.3   Pests

Rape shares a community of pest species with the vegetable brassicas which are grown throughout Europe as both commercial and domestic crops. The significance of individual pest species differs between crops. Pests of importance on oilseed rape in the EEC are:

Blossom beetles
  Meligethes spp.
Pod pests

Ceutorhynchus assimilis	Cabbage seed weevil
Dasineura brassicae	Brassica pod midge

Stem and leaf pests

Ceutorhynchus quadridens	Cabbage stem weevil
Psylloides chrysocephala	Cabbage stem flea beetle
Brevicoryne brassicae	Cabbage aphid
Delia radicum	Cabbage root fly

Nematodes

Heterodera schachtii	Beet cyst nematode
Pratylenchus sp.	Root lesion nematodes

5.3.1   Blossom beetles

Overwintered adult blossom beetles lay their eggs in unopened buds which then do not open, but in winter rape there is considerable compensation for the lost buds. Spring rape is less able to compensate and the pest is proportionately more damaging. Insecticides must not be applied during flowering because of the risk to pollinating insects.

5.3.2   Pod pests

These are generally the most damaging of the pests attacking rape, and the seed weevil is the more important because the pod midge can lay its eggs only in the developing pods through holes made by weevils. The weevils overwinter as adults and enter the crop to lay eggs in young pods. Each larva destroys 3–6 seeds before leaving the pod to pupate. Unlike the weevils, the pod midge can complete two generations on a crop, but the two pests are linked as the midge needs the weevils' feeding punctures to lay its eggs.

78

5.3.3 Stem and leaf pests

Unlike the foregoing pests that directly reduce seed production, the other pests weaken the plant, causing lodging and reduced yield potential. For example, attack by flea beetle larvae increases the incidence of canker due to the fungus *Leptosphaeria maculans*.

Ceutorhynchus quadridens causes serious damage only to spring-sown rape as the eggs are laid whilst the crop is still susceptible to weakening by the larvae which tunnel down leaf veins and stalks into the stem. A related weevil, *C. pictitarsis*, the rape winter stem weevil, is at present quite rare in England, but attacks autumn-sown rape in the same way and overwinters as larvae in the plants. Adults of the cabbage stem flea beetle also lay their eggs in autumn, in the soil near young rape plants. During autumn and winter the larvae find and enter winter rape plants via the leaves and also mine in the stems. Rape sown early in autumn may also be at risk from cabbage root fly attack, although little damage is done in spring.

The cabbage aphid is a sporadic pest, rarely causing significant loss to rape, although it is important on vegetables.

5.3.4 Nematodes

Heterodera schactii has a wide host range including many other crops and weeds as well as oilseed rape. It is likely to increase in importance in the future if arable rotations include a greater proportion of susceptible crops. Suitable rotations can effectively minimise the yield losses caused by this nematode, for the cysts hatch freely in the presence of water and the decline rate in the absence of hosts is more than 60 per cent per annum.

5.3.5 Resistance to pests

Much is known of the deleterious effects on vertebrates of volatile isothiocyanates and the glucosinolates from which they are derived, but relatively little about their effects on insect herbivores not adapted to feeding on *Brassica* species. These substances are key factors for host–plant recognition in the feeding and egg-laying behaviour of the adapted insects belonging to the 'brassica-specialist' community. For example, larvae of the cabbage white butterfly, *Pieris brassicae*, possess chemoreceptors that are specific detectors of glucosinolates. The cabbage aphid, *Brevicoryne brassicae*, will feed on faba bean which is not a host plant in nature if the beans are infiltrated with the glucosinolate sinigrin. The growth rate of *B. brassicae* increases as the allylisothiocyanate content of the plant increases, but the growth of *Myzus persicae*, a species not specifically adapted to brassicas, declines.

Egg-laying by cabbage root flies is promoted by isothiocyanates released by growing plants. Variation in release rates as plants develop can therefore be an important factor in resistance. In radishes, the relative resistance of varieties depends on the age of the plants, with the numbers of root-fly eggs being correlated with the amounts of volatile glucosinolates in the plants. Little is known about resistance to the beetle species, but their responses seem likely to depend on glucosinolate content and metabolism. Reduction of glucosinolate contents in new varieties is unlikely to be sufficient to avoid the adapted pests, but may expose the crop to more attack by pests like *M. persicae*.

Resistance to cabbage aphid occurs in forage rape, but is only effective in vegetative plants since flowering plants are fully susceptible. In New Zealand, adapted aphid biotypes are known that attack resistant varieties. Studies on resistance to this aphid in Brussels sprouts have revealed a number of differing resistant varieties, but there is also variation in aphid populations from different parts of England such that no variety is resistant to all aphids. Resistance of both sprouts and kale to aphids has been associated with glossy (non-glaucous) leaves.

5.4   Seed quality

Rapeseed oil is more valuable than rapeseed meal and so varieties with a high oil content are sought. As rape seed is comprised mainly of oil and protein, high oil content is associated with a low protein content. The other main constituent of seed, fibre, is present in smaller amounts in yellow seeds, which have a thin testa coat. Reduction of the fibre content brings about a concomitant increase in the oil and protein content, though the scope for improvement is limited. Yellow seed coats are found in *Brassica campestris*, *B. juncea* and *B. carinata* but not in rape (*B. napus*) or *B. oleracea*. Attempts are now being made to obtain yellow seed coats in *B. oleracea* by crossing it with *B. carinata*. If successful, the character should also be transferable to oilseed rape. Constituents in the seed coat of normal varieties may reduce intake when the meal is fed to animals.

The fatty acid composition of some major vegetable oils is shown in Table 5.2. Rapeseed oil from traditional winter varieties differs from the other major vegetable oils in that of its fatty acids 45–55 per cent are in the form of erucic acid (22:1). Table 5.3 shows the fatty acid composition of traditional and zero erucic acid varieties and the composition considered ideal for the two main uses of rape oil, the manufacture of margarine and frying. The fatty acids occur in the form of triglycerides. The constituent fatty acids affect the physical and chemical properties of the oil and its suitability for frying and margarine manufacture. In addition the fatty acid composition is considered to be of importance for nutrition and health.

Erucic acid is undesirable because margarine made from oil from traditional varieties will not spread when taken from domestic refrigerators. More importantly, erucic acid is deposited in the heart muscles of experimental animals when they are fed with diets rich in rapeseed oil. This condition, lipidosis, can be followed by the appearance of myocardial lesions. Although for man very much lower amounts of erucic acid are contained in normal diets than in those fed in the animal experiments, it is poorly digested by man and is deemed undesirable. EEC regulations now limit the content of erucic acid in vegetable oils to less than 5 per cent. Fortunately, erucic acid is at the end of a

Table 5.2 *Fatty acids in some vegetable oils (percentage by weight of total fatty acid)*

Source	Myristic (14:0)	Palmitic (16:0)	Stearic (18:0)	Arachidic (20:0)	Oleic (18:1)	Linoleic (18:2)	Linolenic (18:3)
Soyabean	—	9	2	1	32	53	3
Palm	2	42	4	—	42	10	0
Rape*	—	1	1	—	22	22	3
Sunflower	—	5	2	1	35	57	0
Cotton seed	1	21	2	1	25	50	0
Groundnut	—	8	4	3	55	25	0
Olive	1	9	1	1	80	8	0

* Traditional varieties contain *c.* 45 per cent erucic (22:1) and *c.* 10 per cent eicosenoic (20:1) acids.

Table 5.3 *Actual and desired fatty acid composition of rape seed oil (weight of each fatty acid as a percentage of the total fatty acid)*

Fatty acid	Traditional varieties	Zero erucic acid varieties	Desired composition	
			Margarine	Frying
Palmitic	4	5	12	5
Oleic	15	62	28	80
Linoleic	14	22	60	15
Linolenic	9	10	0	0
Eicosenoic	10	1	0	0
Erucic	48	0	0	0

biosynthetic pathway (Fig. 5.1). In 1961 a variant in the German forage rape variety Liho was found in which the synthesis of both eicosenoic and erucic acids is much reduced. Synthesis of these acids is controlled by two multiple allelic loci. By crossing with Liho and selection, varieties of spring rape were produced which had very low levels of erucic acid. The character was transferred to the winter crop and so all the varieties now grown have a very low erucic acid content. Jet Neuf, a variety widely grown in Europe which lacks erucic acid, has a fatty acid composition, expressed as percentages of total fatty acid as follows: palmitic, 5; oleic, 62; linoleic, 22; linolenic, 9; eicosenoic, 3. The composition varies a little with the environment.

For the manufacture of margarine, oil with the following fatty acid content is required: linoleic (18:2), 60 per cent; palmitic (16:0), 12 per cent; oleic (18:1), 28 per cent; and linolenic acid (18:3), none (Table 5.3). Linolenic acid is unstable and is readily oxidised to unpleasant-smelling substances. For frying, oil with 80 per cent oleic acid and no linolenic acid is required. Breeders believe that their present programme will enable varieties of rape to be produced with higher palmitic and oleic acid than at present but they do not envisage being able to increase the linoleic content beyond 40–45 per cent or decrease the linolenic content to lower than about 3 per cent. It will take about ten years to breed varieties with the fatty acid composition required for the two major uses. It is important, therefore, to explore ways of achieving these objectives more completely and more rapidly.

For the manufacture of margarine, oil with a low sulphur content is needed. This is so because sulphur poisons the nickel catalyst used in the hydrogenation process. Some of the sulphur in oil is in the form of dissolved glucosinolates. The glucosinolates are present mainly in the

Fig. 5.1 Biosynthetic pathways for fatty acids.

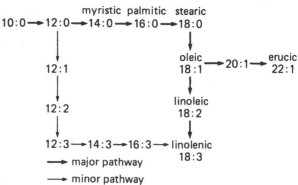

meal and it is desired to reduce their amounts in the meal (see below). To the extent that this objective is achieved, the sulphur content of the oil will be reduced concomitantly.

Two other constituents of the oil are undesirable. Phospholipids are partly responsible for gumminess and constitute about 3.5 per cent of the oil. They can be removed chemically, but the process is expensive. There appears to be little genetic variation in phospholipid content in rape, though sunflower oil contains only 0.2–0.7 per cent. Because phospholipids are essential constituents of living cells, it is unlikely that they can be eliminated entirely from oil. It is likely that their concentration in vegetable oils is partly a function of the high percentage of oil in the seed. However, selection for reduced phospholipid content in oil may be effective if their composition is changed so that they are less readily extracted during crushing.

Rapeseed meal has a better balanced amino acid composition than many other seed meals, including soyabean meal. However, meal from varieties presently grown contains a number of other compounds which make the meal unsuitable for feeding to some animals. One group of undesirable compounds are glucosinolates, which have the generic structure shown below:

$$R—C{\overset{\displaystyle S—[glucose]}{\underset{\displaystyle N—O—SO_2OH}{}}}$$

There are at least two classes of these compounds, one in which R is an aliphatic chain and the other in which it is a substituted indole. In the meal, glucosinolates are hydrolysed by the enzyme myrosinase to give glucose and goitrogenic aglycones (isothiocyanates, nitriles and thiocyanates), which have a characteristic pungent odour and 'biting' taste. Some of the aglycones from glucosinolates which are goitrogenic in non-ruminants are competitive inhibitors of iodine uptake by the thyroid gland and so their effects can be countered by supplementing the iodine content of the diet. Others, notably goitrin, inhibit the synthesis of thyroid hormone and iodine supplementation is ineffective. However, it is the bitter taste of the aglycones that reduces the palatability of the meal to animals, so reducing intake and hence their growth rates. Partly as a consequence, the inclusion of rape meal in feed compounds is limited to about 10 per cent for cattle, 5 per cent for pigs and 2.5 per cent for broilers. Other factors (see below) preclude the use of rape meal in rations for laying hens.

An important objective in the improvement of rape is therefore to breed for a lower glucosinolate content. The Polish variety Bronowski is a major source of genes for low glucosinolate content, having only about 10 per cent of the content of normal European varieties. The maternal parent determines the glucosinolate content of the seed, the pods being the main sites of synthesis of the compounds which are present in the seed. At least three partially recessive genes determine the glucosinolate content of the seed.

It is likely that the types of glucosinolates as well as their total amount can be modified by breeding and possibly by novel methods of genetic manipulation. However, the types of glucosinolate found in *B. napus* varieties do not vary much. It is suspected, but not proved, that glucosinolates play a role in determining the interactions between the rape plant and insect pests and fungi. If so, it may be desirable to retain or increase the concentrations of individual glucosinolates in vegetative plant parts while reducing concentrations in the seed. This may prove a difficult task related to gene expression in different tissues. Also, the extent to which glucosinolates or their aglycones are translocated from the vegetative organs to the seed is presently unknown.

The inclusion of rape meal in cattle rations could be doubled with low glucosinolate meal, but it will still not be suitable for feeding to laying poultry. The fishy taint which can develop in the eggs of some breeds of brown egg layers is not itself related to glucosinolates but is due to trimethylamine, the precursor of which is sinapine, present at levels of 1–2 per cent in rapeseed meal. A genetic defect is associated with the taint in such hens: a reduced ability of the kidney and liver to oxidise trimethylamine to water-soluble trimethylamine oxide, which is then excreted. This reduced capacity is further aggravated by goitrin, which appears directly to inhibit hepatic trimethylamine oxidase activity. As the level of progoitrin in low glucosinolate cultivars is in excess of that required to inhibit trimethylamine oxidation and the variation in sinapine content in *B. napus* is small, the most appropriate solution to the problem seems to be to breed poultry flocks that produce brown eggs but do not have this impaired ability.

5.5   Molecular biology and the improvement of oilseed rape
5.5.1   Improvement of yield
    Most of the limitations to yield discussed in Section 5.1 would be best investigated by carrying out selection experiments, exploiting the variability already available to breeders. This experimentation, which could be done as part of a breeding programme, would not be aided by molecular biology.

Moricandia arvensis, a member of the Brassiceae, has an 'inter-mediate', C3–C4 photosynthetic carbon metabolism as indicated by the presence of chloroplasts and numerous mitochondria in the bundle sheath cells, by the presence in the leaf tissue of considerable phosphoenol pyruvate (PEP) carboxylase activity and by a carbon dioxide compensation point intermediate between that of C3 and C4 species. This feature could be valuable in oilseed rape, by causing a reduction of photorespiration and hence an increase in net photosynthesis, although it may turn out to be a feature that confers fitness to temperature and light environments not usually encountered by oilseed rape. If further study shows that it might be useful, and the character cannot readily be transferred by conventional genetic means, molecular biology could aid the incorporation of the character into oilseed rape.

A more speculative exercise would be to isolate photosynthesis genes from, say, wheat, and transfer them to oilseed rape. Dicots, as a class, tend to have lower light-saturated rates of photosynthesis than do the Graminae. If the biochemical and genetic basis for the differences could be established, particular target genes for transfer could be defined. This method of experimentation would also allow the effects of related genes from different Graminae to be tested. A major difficulty would be that of eliminating or turning off in the oilseed rape recipient the homologous genes, though the effects of added wheat genes could be assessed. It is not likely to be a simple matter, having inserted the required gene and eliminated its counterpart in the host, to get it expressed in the donor in the right cells and organs at the right time and in the right amount. This approach, therefore, whilst it should be successful in principle, would be a considerable and expensive task. Target genes would include those coding for the light harvesting chlorophyll protein of photosystem II and for the small subunit of Rubisco, if these differed in the donor and recipient species.

It has been noted (Section 5.1) that the resistance to atrazine which has been transferred into oilseed rape appears to be associated with reduced vigour. It would be beneficial to incorporate resistance to another herbicide (not a triazine), preferably of a kind determined by a nuclear gene or genes. If the herbicide were also used for weed control in cereals, it should be possible to identify and transfer the resistance from cereals to oilseed rape. Alternatively, if resistance was determined by genes expressed in isolated protoplasts or callus cultures, mass selection for resistance could be undertaken in the laboratory. A further alternative source of resistance, relying on degradation, would be microorganisms, and transfer of genes from them to oilseed rape should now be possible. The choice of a suitable herbicide would depend on economic, biological and environmental considerations but the substituted ureas and the nitriles (examples: chlortoluron and ioxynil) are possible candidates. Both classes of herbicides affect the light reactions of photosynthesis, however. Resistance, or tolerance, may depend on chloroplast-encoded genes and may only be expressed in green cells. As alternatives, the carbamates and imidazolinones might be considered. It should be borne in mind that the development of new herbicides is a continuing process and that a more effective and cheaper selective herbicide for the crop could be developed, thus avoiding the need for herbicide tolerance.

In oilseed rape as in other Brassicas, there is a self-incompatibility system which prevents self-pollination, making the crop out-breeding. In oilseed rape there are few alleles available and it would be desirable to have a wide choice so that they could be exploited in producing hybrids (as has been done so successfully with kale, *B. oleracea*). The molecular basis of sporophytic self-incompatibility of the kind present in the Brassicas is beginning to be understood and a cDNA has been prepared in the USA which can be used to detect homologous sequences in preparations of DNA from seedlings. There is thus the possibility that from an extension of this work, other cDNAs could be prepared which would enable particular alleles to be detected in seedlings, greatly aiding the exploitation of self-incompatibility in the species.

5.5.2   Improvement of resistance to disease
Improved seedling resistance to stem canker (*Leptosphaeria maculans*) would be valuable. Seedling resistance is present in the B genome of *B. juncea* and hence cannot be transferred to *B. napus* by crossing. The basis of the resistance of *B. juncea* is not known. If further research showed that it had a simple molecular basis and the genes could be identified, it should be possible to transfer them to oilseed rape.

Useful resistance to the diseases of the brassicas, clubroot (*Plasmodiophora brassicae*) and stem rot (*Sclerotinia sclerotiorum*) has proved difficult to introduce into cultivated species, and despite much research, the mechanisms of resistance are not known. Application of the techniques of molecular biology could help to provide this information and show whether and how resistance could be improved. This is a considerable challenge to pathologists and molecular biologists.

5.5.3   Improvement of resistance to pests
As in the case of the shoot fly species attacking wheat, the pests of oilseed rape form a group of related insects, weevils and flea beetles,

that, taken separately, hardly warrant the expenditure of effort on breeding for resistance at the present. However, if the techniques of molecular biology could be used to introduce a form of resistance that controlled all these pests, the effort could be worthwhile.

Two approaches seem most likely to be realistic. First, if variants of *Bacillus thuringiensis* toxins with high activity against beetles could be identified, the genes could be transferred to rape to control most of the current pests (and they could also be exploited in other crops, e.g. potatoes). The alternative approach likely to give a general degree of resistance would be the disruption of the 'brassica-adaptation' that is characteristic of these insects. This might be achieved by introduction of an alien secondary metabolite giving an appropriate scent or flavour.

Otherwise, the focus should be on individual species. It would be worthwhile to evaluate potential sources of resistance, for example the oviposition-deterrent pheromone of the white butterfly *Pieris brassicae*, with regard to all brassica crops affected by the pest, for, if one of the crops can be transformed, probably all can.

5.5.4 Improvement of seed quality

Linolenic acid, constituting about 8 per cent of the fatty acids, is undesirable. As it is at the end of a biosynthetic pathway (Fig. 5.1) it might be possible to induce mutations lacking the capacity to synthesise the acid. However, as it is synthesised by two routes, complete elimination would require the loss of at least two enzymes. Linolenic acid is an essential constituent of the thylakoid membranes of chloroplasts. As rape seed cotyledons are green before maturity and after germination, it is possible that a reduced linolenic acid content in the seed would be associated with reduced fitness.

As already noted, the feeding value of the meal would be greatly improved by the elimination of glucosinolates. Using low glucosinolate varieties as a starting point, mutations induced either chemically or by irradiation could be sought which had reduced glucosinolate content. There are rapid methods for analysis for glucosinolate content and so it should not be too difficult to identify relevant mutants. However, as with all mutants, there would be the risk of undesirable pleiotropic effects and these should be anticipated and searched for at an early stage in testing.

6
Faba beans

Faba beans (*Vicia faba*), or field beans, are a traditional crop in Europe used as field beans (*V. faba minor* and *equina*) for animal feeding and as broad beans (*V. faba major*) for direct human consumption. They are also an important crop in Mediterranean countries as a source of protein for human consumption. In the EEC the area planted with faba beans and their production gradually declined until the late 1970s when the EEC fixed a target price for faba beans. Since then production has increased and is now about 0.35 Mt a^{-1}. The EEC policy of encouraging the production of faba beans (and other grain legumes) was a response to the doubling of fish meal prices and the trebling of soyabean prices in 1972 and the embargo subsequently (1973) imposed by the USA on the export of soyabeans.

Of the total oilseed cake and meal (i.e. feeding-stuffs ingredients rich in plant protein) used in making animal feeds in the EEC in 1980 (19 Mt), about 60 per cent was imported. Much of the importation was soyabean (as seed or meal), with smaller quantities of cotton seed, copra, groundnut, palm kernel and other meal. In terms of the protein from oilseeds, the percentage imported was considerably more, being 92 per cent in 1982. Maize gluten is now widely available and is a partial substitute for the traditional oilseed cakes and meals. In addition to high protein meals from plants, EEC countries also imported about 20 Mt of animal and fish meals in 1980 for inclusion in feeding-stuffs. Plant protein meals from crops grown in the EEC could provide a partial substitute for the imported plant and animal meals, provided its composition and production costs were favourable.

Whilst exact statistics are difficult to obtain it is clear that there is scope for the EEC countries to produce much more high-protein meal for animal feeding. Species of potential interest include oilseed rape, faba beans, peas and lupins, which are suitable for northern Europe, and sunflower, soyabeans and lupins, which are suitable for southern Europe. As is often pointed out, faba beans and peas are alternative crops to wheat and barley so any increase in the area sown with these pulses would reduce the area sown with cereals, and so reduce the present surplus of cereals. What is needed therefore, are varieties and production systems for these crops which would enable them to be grown more easily and more profitably than at present.

6.1   Biological limitations to yield

The potential yield of faba beans is difficult to predict from existing knowledge. Unlike wheat, vegetative growth continues during seed growth, though, because of losses from lower parts of the plant, there may be a net loss in the weight of the vegetative organs. Flowering takes place over a period of at least 2–3 weeks. The photosynthetic characteristics of the leaves are very imperfectly understood.

A further uncertainty in the calculation of a potential yield from physiological principles is the energy cost of nitrogen fixation. Assuming that 12 moles of carbon dioxide are required per mole of nitrogen fixed in the nodules, a crop of average composition (30 per cent protein in the seed and a harvest index of 50 per cent) would yield at most 85 per cent of the weight of grain of an otherwise similar non-legume grain crop for the same photosynthetic input. In other words, there is approximately a 15 per cent yield penalty for nitrogen fixation. In view of these uncertainties, one can probably do no better than to accept the record yield (about 9 t ha^{-1}) as a reasonable approximation, though probably an underestimate, of the potential yield.

The average yield of faba beans in the EC countries was 1.8 t ha^{-1} in 1983. This was much lower than the average yield of wheat (4.5 t ha^{-1}). Further, the variability in the yield of faba beans is higher than that of cereals. Taking data from the UK, coefficients of variation of annual average yields of faba beans, wheat and barley are 11.7, 7.3 and 5.9 per cent respectively. The variability of faba bean yields in trials appears to be close to that of wheat and barley and so it may be that the reputation for high variability of faba beans is due more to the skill and attention devoted to the crop than an inherent feature of the species. The balance of opinion is that the crop is more difficult to grow than cereals, is more subject to the effects of adverse environments, especially drought and excessive rainfall, and to pests and diseases. In addition, the crop is not reliably self-pollinating, as are cereals, and so a good level of fertilisation depends on the presence during the flowering period of a sufficient number of pollinating insects (bees, including certain species of *Bombus*). Poor pollination is considered to limit seed set during the early part of the flowering period of winter beans in the UK and generally in sourthern Europe. The problem may be aggravated by growing the crop in large fields, where *Bombus* populations are too small to ensure pollination.

In northern Europe except France and the UK, the winters are too cold and only spring varieties of faba bean are grown. In the UK and France, winter and spring varieties are grown and in the UK the winter varieties predominate at present. Occasionally (about once in fifteen years), the winters are sufficiently cold in England to cause appreciable winter kill.

Yields of faba beans in trials (4.5 t ha^{-1} in a series of 22 trials with spring beans in the EC countries during 1977–79), whilst much greater than farmers' yields, are modest compared with farmers' yields of cereals or cereal yields in trials. The highest yields of beans in the same set of 22 bean trials were close to 9 t ha^{-1}. Yields of spring wheat rarely approach this level though the highest yields of winter wheat are close to 12 t ha^{-1}. In the EEC, average bean yields on farms are only about 20 per cent of the maximum, while average wheat yields are 42 and 48 per cent (spring and winter crops, respectively) of an assumed maximum of 12 t ha^{-1}. Taken with the evidence of the greater variability of faba bean than cereal yields, these figures suggest that yields of faba beans are indeed more limited by environmental hazards, pests and diseases than are those of cereals, possibly because farmers do not give the crop the degree of attention needed to minimise such environmental and biotic hazards. If this analysis is correct, in breeding new varieties, attention should be given to characters which will enable yields to approach more closely the present potential. This may be easier and more rewarding than striving to increase the potential yield.

When water is freely available the vegetative apex continues to grow, resulting in the production of more leaves and stem but no commensurate increase in seed yield. Such vegetative growth is regarded as 'excessive'. On the other hand, faba beans appear to be quite sensitive to drought, which is probably an equally important cause of loss of yield. Furthermore, it restricts the range of soils suitable for growing the crop to heavy clays which have large reserves of available water. Drought more often affects the spring bean crop, because it flowers later than the winter crop.

In an attempt to reduce the variable growth and yield caused by fluctuations in water supply, breeders are incorporating the terminal inflorescence (*ti*) or 'topless' mutant into their breeding material. The mutant prevents the excessive growth in length of the stems which can occur when water is freely available from the beginning of June, though, as if to 'compensate', more tillers and nodal branches are formed, especially at low density. Tillers so formed do not usually bear seed and so are 'wasteful'. This response may be presumed to be under genetic control and so selection against late tiller and nodal branch production should overcome the problem. Advanced breeding lines incorporating the *ti* mutant yield only 3–6 per cent less seed than conventional varieties, but it is not yet known whether the *ti* lines are less variable over locations and years. Given the relatively short time that the gene has been used, the prospects seem good for the production of high-yielding, stable varieties with the *ti* mutant.

The *ti* mutant does not offer any prospect of increasing the drought resistance of the crop. Many legumes from subtropical regions, including crop species (notably chickpea, *Cicer arietinum*) have good resistance to drought. In the long term, therefore, there may be prospects for lessening the sensitivity of faba beans to drought by transfer of appropriate genes from these species. Genes which modify stomatal behaviour, osmoregulation and leaf abscission would be the most obvious targets for identification and transfer.

Orobanche (broomrape) is a serious weed of faba beans in Mediterranean countries, where it is the major factor limiting production. Glyphosate at low rates selectively kills the parasite but the dose is critical. However, it can be used to kill the inflorescence (i.e. the above-ground parts of the weed) and prevent seeding. Genetic resistance to *Orobanche* is reported to exist in the exotic cultivar Giza 402 and might be suitable for incorporation into other varieties.

6.2   Diseases

In addition to the diseases caused by *Botrytis fabae* and *Ascochyta fabae* a number of other diseases are of potential importance. These include those caused by *Uromyces* sp., *Fusarium* sp., *Sclerotinia* and at least five viruses. None of the viruses is considered to be of major importance at present.

6.2.1   *Leaf spot* Ascochyta fabae

This is mainly spread through infected seed, hence a seed certification scheme operates (in the UK). However, there is evidence for infection via 'volunteer' plants and it is desirable to have an adequate level of resistance in commercial cultivars. There is varietal variation in resistance and a good level of resistance in one of the inbred winter lines from the breeding programme at the Plant Breeding Institute, but there is little or no information on the genetics of resistance, cultivar specificity of pathogen populations, or mechanisms of resistance. The disease is of considerable concern to breeders of *V. faba* in the Mediterranean countries.

Ascochyta pisi, which attacks peas, is known to liberate cell-wall degrading enzymes; a large proportion (*c.* 78%) of the activity of which is attributable to an endopolygalacturonase (EPG). This enzyme can reproduce disease symptoms when applied to healthy tissues and appears to be the first enzyme produced on infection by the fungus. A pea cultivar has been shown to produce a soluble protein which inhibits EPG and hence this may offer a form of resistance to infection which could be transferred to faba beans.

6.2.2   Chocolate spot (Botrytis fabae)

This disease occurs on faba beans in the UK in most seasons but only causes severe damage and yield loss in years particularly favourable to its development (wet conditions in June–July). The disease is a '*compound*' one – both *Botrytis fabae* and *B. cinerea* and associated bacterial species may be present in the lesions which may occur on leaves, stems and sometimes on pods and seed coats. Infection is mainly by conidia liberated from conidiophores produced on decaying vegetation. Hence, the fungi can live saprophytically. Infection is favoured by high humidity, though is restricted by high temperatures.

Good resistance in *V. faba* to *Botrytis fabae* has been derived in breeding programmes from several sources but the best source is a line from Ecuador. Higher levels of resistance exist in other *Vicia* species but incompatibility prevents transfer of these genes.

The genetics of resistance to chocolate spot in *V. faba* are imperfectly understood. Early work suggested that resistance was controlled by two genes. Resistance expressed through the hypersensitive response (HR), in which infection is restricted by cell necrosis, appears to be under the control of a number of dominant genes, which may be partially additive in action.

B. fabae is considered the primary parasitic organism in chocolate spot with *B. cinerea* being secondary and much less virulent. However, it has been claimed that *B. cinerea* can sometimes produce the aggressive phase of the disease. Comparisons have been made between these two species, and there is considerable information on the infection process and resistance mechanisms in *V. faba*. *B. fabae* produces a polygalacturonase and infection may be aided by associated bacteria which produce a related cell-wall degrading enzyme, polygalacturonic acid trans-

92

eliminase. *V. faba* produces phytoalexins in response to infection with *Botrytis*. Six chemically related substances are known, all of which are more active against *B. cinerea* than *B. fabae*. Evidence that specific phytoalexins provide a mechanism of resistance, at least to *B. cinerea*, was initially compelling. Cotyledons are resistant to infection by both *B. fabae* and *B. cinerea* and accumulate the highest levels of the phytoalexin wyerone of any tissue. Although some work has suggested phytoalexin accumulation to be a basis for differential infectivity by the two fungi, later work has cast doubt on this and it has not been established whether there is any relationship between the aggressive phase of *B. fabae* infection and phytoalexin production. *B. fabae* more readily penetrates and kills epidermal cells during initial infection than do other *Botrytis* species, but differences occur before accumulation of phytoalexins in the tissue. Specificity may partly depend on the ability of the fungus to enter and rapidly kill epidermal cells before they can produce phytoalexins. This in turn may depend on the formation of cell wall degrading enzymes by the fungus or its associated bacteria. In addition, however, the ability to resist phytoalexins may also contribute to the pathogenicity of *B. fabae*. Uncertainties include the degree to which the bean phytoalexins are compartmentalised (and hence the effective concentration present in the tissue) and the conditions pertaining during infection, as in *in vivo* tests pH and media composition can have marked effects on antifungal activity. In addition, strains of both *B. cinerea* and *B. fabae* are known to differ in their susceptibility to wyerone. There appears to have been no investigation of whether *V. faba* varieties differ in their ability to accumulate phytoalexins or whether differences in susceptibility to infection are related to phytoalexin production.

6.3   Pests

Pests of importance in the culture of beans include:

Aphis fabae	Black bean aphid
Sitona lineatus	Bean weevil
Apion vorax	
Ditylenchus dipsaci	Stem nematode

6.3.1   Aphis fabae

The aphid is the most important of the pests, and its biology has been extensively studied. It tends to occur in damaging numbers in alternate years. The aphid will colonise a wide range of plants as well as beans and sugar beet, which are its normal crop hosts. Spring-sown crops that are at an earlier growth stage at the time of the spring aphid migration than autumn-sown crops are more severely damaged. Yield is lost due to weakening of plants by the feeding of dense aphid colonies. The feeding behaviour of *A. fabae* is adapted to rapid exploitation of the richest food sources in the plant; it extracts more phloem sap and excretes more nitrogen than aphids of other species feeding on similar plants. Unchecked infestations in sparse bean crops may reduce yields to less than half that of sprayed crops.

In Britain, *A. fabae* normally overwinters as eggs laid by sexual females on spindle (*Euonymus*) bushes, and the severity of crop infestations can be forecast by observing the numbers of eggs on these plants. In southern Europe, the sexual phase is less usual and the parthenogenetic summer forms normally persist through the winter. Therefore, aphid clones adapted to particular plant genotypes are more likely to reappear in successive seasons than in northern Europe.

Resistance to bean aphid occurs only at low levels in cultivated beans, but varietal differences have been reported several times. This resistance seems to implicate not only the morphology or biochemistry of the plant, but also its physiological functioning, particularly with respect to the phloem which is the aphids' food source. Some resistant varieties differ from the most susceptible ones in their pattern of growth, as plants with dispersed growth offer less favourable feeding sites than plants with few vigorous growth points. In other cases, the aphids aggregate to feed on veins that are larger on resistant and relatively smaller on susceptible plants. It seems likely that new varieties with determinate growth will be much less susceptible to aphid attack, especially when determinate varieties suitable for autumn sowing are developed.

Higher levels of resistance occur in wild *Vicia* species related to faba beans, and some of these are virtually immune to *A. fabae*. *V. johannis* is the most resistant of the species closely related to faba beans. Resistance is not transferable from these species by conventional breeding techniques because hybrids with *V. faba* cannot be made. Susceptibility to three species of aphids amongst these beans is positively associated with the degree of domestication, and resistance to the number and concentration of non-protein amino acids. The only non-protein amino acid in faba bean varieties is L-DOPA, a feeding stimulant for *A. fabae*. However, L-DOPA is absent from tannin-free faba bean genotypes, but they are not more resistant to *A. fabae*.

6.3.2 Sitona and Apion

Adult *Sitona* weevils cause very conspicuous damage by chewing half circles from leaf edges, especially in spring when they migrate to bean fields after hibernation. However, the loss of leaf area very rarely affects yield. The larvae of *S. lineatus* feed in spring and summer below ground on the root nodules, presumably reducing nitrogen fixation, and the presence of many larvae may cause significant losses of yield. Attacked crops become yellowish and yields may be reduced by up to 0.5 t ha^{-1}. *Sitona* is also important as a vector of virus disease, but another weevil, *Apion vorax*, is much more efficient in transmitting broad bean mosaic virus and broad bean stain virus. The larvae of *A. vorax* develop in the flowers of the bean, but the species also lives on wild vetches.

Little is known about varietal susceptibility to attack by these weevils. There are several *Sitona* species that feed on different legumes in Europe, but faba beans are not hosts to the larvae of most, indicating that host selection is important for these insects. It seems likely, therefore, that resistant bean genotypes could be found.

6.3.3   *Ditylenchus dipsaci*

Stem nematodes exist as a number of races with differing host-plant ranges. Two races are important for beans, the oat race, whose wide host range includes oats, rye, maize, sugar beet, peas and many weeds, and also a tetraploid, giant race that appears specific to faba beans. The giant race of *D. dipsaci* nematodes weakens and distorts infected plants, reducing their yield. Chemical control is difficult and cultural preventative procedures, including rotation, straw burning and weed control, are the farmers' best defence. The races that attack beans may enter the seeds as they develop, and beans are therefore a significant crop in relation to this pest, as infected seed can cause infestations.

Resistance to stem nematode is known in other leguminous crops, such as red clover and lucerne, where resistant varieties and lines were bred, but this has not been attempted with *V. faba*. There is evidence of variation in the susceptibility of bean varieties and a land race from Morocco, Souk el Arba du Rharb, is claimed to be resistant. The durability of resistance is difficult to predict as the variation of host responses and the taxonomic status of the nematode races is not well known.

6.4   Seed quality

The main factor governing the use of faba beans for animal feeding in the EC countries is price, rather than quality. Unless the costs of production can be decreased (e.g. by raising and stabilising yields) faba beans will be unlikely to be competitive with imported soya meal or maize gluten. Thus, research to improve quality is of secondary importance to that directed to decreasing the unit cost of production. The main factor which would limit the acceptibility of faba beans for animal feeding is the presence of antinutritive factors. These are discussed below.

Because faba beans are a minor crop, at present of little importance for animal feeding, the biochemical and genetic basis of factors affecting quality is not well understood. In the absence of this information it is impossible to assess with confidence the scope for improving quality either by conventional or novel methods of breeding.

The main difference between the gross composition of soyabeans and faba beans is the lower protein content of the latter. Protein content could be increased by selection and although it is possible that, as with wheat, there would be a pleiotropic reduction in yield, it may not be so difficult to overcome in beans.

The storage proteins of faba beans, like those of other legumes, are the globulins legumin and vicilin. Typically the ratio of legumin to vicilin is about 2.0, but there is genetic variation in the ratio and values between 1.6 and 2.5 are reported in the literature. The ratio also varies with environment, being higher in environments which increase the total protein content of the seed. Legumin has a molecular weight of about 330 kD. It is made up of six subunits, each comprising an acidic polypeptide of molecular weight c. 36 kD, and a basic one of molecular weight c. 20 kD. The two polypeptides of a subunit are linked by a disulphide bond. There is heterogeneity within each kind of polypeptide as assessed by N-terminal amino acid sequences and by gel electrophoresis. In addition the subunits belong to two structural types: type A contain methionine, whereas type B do not. As this amino acid is nutritionally limiting for monogastric animals there may be scope for overcoming this limitation by increasing the abundance of the type A subunits. Of the two other sulphur-containing amino acids, there is little cysteine in legumin but variable amounts of lysine, averaging 4 mol per cent, depending on the subunit type and charge.

Vicilin has a molecular weight of about 186 kD and is glycosylated. It is probably a mixture of several protein species. It comprises three major subunits of approximately equal molecular weight and there is some genetic heterogeneity at the subunit level in *Vicia*. Like legumin, vicilin also has a high content of asparagine and glutamine and is relatively deficient in methionine. However, its lysine content is much greater than that of legumin.

Recent studies on the legumin and vicilin storage proteins in soyabean show that there is a considerable degree of homology between the two proteins. The major difference between the proteins is that the legumin subunits have a polypeptide sequence inserted between the second and third domain of the protein. This insert is of variable length and is rich in aspartic and glutamic acids and gives the acidic character to the polypeptide. It seems probable that the legumin and vicilin in faba beans

96

are broadly similar to the corresponding proteins in soyabeans. Certainly, a comparative study of these proteins, and of related storage proteins in other species, should reveal the scope for making genetic changes in faba beans.

The variation in the relative proportions of legumin and vicilin, and the variation in the amino acid composition of the legumin subunits, suggest that there is scope for varying the amino acid composition of the total protein in faba beans to make it more suitable as a constituent of feeds for monogastric animals. However, the variation in the ratio of methionine to protein among varieties of faba beans appears to be small. For feeding to ruminants, the protein is too readily degraded and it is not much better than urea as a nitrogen source.

Faba beans contain a number of constituents which reduce feed intake or the digestibility of the protein they contain. These are most significant when the beans are used in feeds for monogastric animals. Chief among them is tannin, contained in the seed coat. Tannin in the seed coat is pleiotropically associated with melanin wing spot on the petals, varieties with white flowers having little or no tannin in their seed coats. Traditionally, faba beans grown for animal feeding are of the high tannin kind and only some varieties grown as vegetables have white petals and low tannin. Feeding trials with pigs in which the diets were arranged to have equal protein and metabolisable energy contents showed that the nitrogen retention was 7.3 g day^{-1} for the low and 6.9 g day^{-1} for the high tannin kinds. In the same trial the nitrogen retention from a soyabean diet (of the same protein and metabolisable energy content) was 8.3 g day^{-1}. This and limited other evidence suggests that low tannin beans would be preferable to normal, high tannin ones. Of course the seed coats can be removed by de-hulling, and this will eliminate the tannin and enhance the feeding value of the bean meal, but at the loss of about 10–15 per cent by weight in yield, though a smaller percentage loss in metabolisable energy. In the field, the establishment of plants from the seed of some tannin-free isogenic lines is poorer than that of their tannin-containing counterparts. It is not yet known whether this is a pleiotropic effect of the absence of tannin from the seed coat, or whether it is a consequence of linkage with other genes which affect germination or seedling vigour. Although not unequivocally established, there do not appear to be any other undesirable pleiotropic effects of low seed coat tannin content on yield, pest or disease resistance, or quality.

Other factors considered to be of minor importance for quality are lectins, phytate and trypsin inhibitors. The limited evidence suggests that these could be reduced by selection. Of more importance when faba beans are fed to poultry is the presence of substances, notably convicine, vicine and the non-protein amino acid L-DOPA, which reduce nutritional value. Selection to remove these substances would be beneficial. As they also induce favism in humans the varieties lacking these substances would be beneficial in countries of the Mediterranean basin where faba beans can form a significant item in the diet and where a greater proportion of the population is susceptible to favism than in northern Europe.

6.5   Molecular biology and the improvement of faba beans

6.5.1   Improvement of yield

It is difficult to identify with certainty characters which, if incorporated into faba beans, would increase yield and decrease yield variability. It seems unlikely that the *ti* gene will have a dramatically beneficial effect, even in the long term, when other plant characters have been appropriately modified. Yield losses resulting from poor pollination, caused by inadequate numbers of *Bombus* bees and their low levels of activity during cool, wet weather, would be reduced if the crop possessed a greater degree of autofertility. The genetic control of autofertility is not understood, but it is probably polygenic, making improvements by molecular biological methods difficult to envisage.

Increased winter hardiness is a possible exception to this as it would enable autumn-sown varieties, with greater yield potential than spring-sown ones, to be grown more widely. Although winter varieties are more likely to escape the effects of drought, improved resistance to drought would probably be beneficial. As noted in Section 6.1, if genes for characters determining drought resistance in other legumes could be identified, these would be candidates for transfer to faba beans. Such characters, probably genetically and physiologically complex, however, include: modified abscisic acid biochemistry, improved osmoregulatory capacity, greater cuticular resistance to water loss and deeper rooting.

Although legumes generally are not difficult subjects for tissue culture, little research on this topic has been carried out with faba beans and a transformation system has not yet been demonstrated.

Breeders believe that it would be beneficial to have means of introducing alien variation to faba beans and possible techniques include protoplast fusion, pollen irradiation and microinjection with foreign DNA. Additionally, any methods which could be used to generate isogenic lines quickly would facilitate character evaluation which is at present very expensive and time consuming.

There is scope for modifying the breeding system of beans to make them more amenable to improvement. Improved cytoplasmic male sterility based on '447' type cytoplasm would make it feasible to produce F_1 hybrids (it has not yet been demonstrated whether chemical hybridising agents are effective in causing male sterility in beans, or would be suitable for F_1 hybrid production). Alternatively, improved autofertility combined with cleistogamy derived from exotic sources (Afghanistan, Pakistan and Ethiopia) would convert faba beans into a self-pollinating crop, making pedigree breeding, so successful in cereals, a practical system. Both these alterations to the breeding system would inevitably take more than a decade to evaluate satisfactorily, and might not be aided by molecular biology. However, if one or other system proved a practical means of producing varieties, it would then be easier to benefit from any gene transfer systems based on molecular biology which might in the meantime have been developed.

The seed of *Orobanche* requires specific germination stimulators from the host in order to become established on it. It is reasonable to suppose that if a step in the biosynthetic pathway of a major stimulator could be eliminated, the plants lacking this step would be resistant to *Orobanche* infestation. Whilst other resistance mechanisms of an active kind could be envisaged, it might be relatively easy to use mutagenesis to create resistance by the former route.

6.5.2 Improvement of resistance to disease

Resistance to *Ascochyta fabae* is required but it is not yet clear whether adequate levels of resistance are present in adapted and exotic races. If they are not, a wider search for resistance will need to be made, and if found, the resistance may be transferable only by molecular biological methods.

Resistance to *Botrytis fabae* occurs in species related to faba beans and work to transfer the resistance to faba beans needs to be initiated. If it is assessed that resistance in faba beans could be obtained by gene deletion or inactivation, mutagenesis, possibly at the protoplast level, and mass screening could provide an easy way of generating resistance to the disease.

6.5.3 Improvement of resistance to pests

Black bean aphid is the major pest, but before devoting further effort to generating new resistant genotypes by molecular biology or other techniques, it may be advisable to await the transfer of the determinate habit to autumn-sown varieties and to assess the severity of the aphid problem on this new model of the faba bean crop. If further modifications are then deemed worthwhile, characters from the wild relatives of the crop, including but not only non-protein amino acid content, may be worth transferring. Protoplast fusion may offer a means of achieving this.

An increase in the area of faba beans grown, which might follow the introduction of new high-yielding varieties, could make *Sitona* a more significant pest than it is now. More work is needed to understand its biology and host-plant relationships. However, if vector-associated genes are constructed for resistance to weevils in rape, it could well be worthwhile introducing them to faba bean also to control *Sitona*.

An alternative method of controlling *Sitona* might be developed by transforming the nodule-bacteria by introduction of an insecticidal molecule. An effective method of distributing and ensuring the establishment of the transformed bacteria in nodules would be necessary.

6.5.4   Improvement of quality

It might be possible to increase the protein content of faba beans by increasing the number of copies of the genes controlling the synthesis of legumin and vicilin, or by increasing the efficiencies in the promoter regions of the genes. However, this approach would only succeed if a limitation to protein content was the synthetic machinery in the seed and not in the supply of amino acids from the vegetative parts of the plant. An alternative would be to reduce starch synthesis in the seed but this might have undesirable side-effects.

Recent evidence suggests that the 7S and 11S (vicilin and legumin in the Leguminosae) proteins have regions which are highly conserved in angiosperms, and other regions which are variable. As the amino acid profiles of the corresponding proteins in other species (e.g. soyabeans and oilseed rape) are more favourable for animal feeding, there is the prospect in the long term of transferring the coding sequences for the variable regions rich in methionine from other species to faba beans. Alternatively, it might be possible to replace the variable regions with synthetic oligonucleotides based on the naturally occurring ones, but with the sequences coding for methionine being replaced for the structurally similar amino acid histidine.

Mutagenesis might be used to eliminate or reduce convicine, L-DOPA and vicine contents.

7
Conclusion

Recombinant DNA technology is a rapidly developing subject which offers many entirely new opportunities for making directed changes to plant genotypes. It may supplement the many techniques already used by breeders to increase the yield, pest and disease resistance and quality of crops, but has not yet been demonstrated to be generally useful. Provided that an appropriate level of funding is maintained, and appropriate collaboration between scientists working at different levels is achieved, the next two decades should permit the new technology to be further developed and properly tried and tested. To facilitate this, it is necessary to identify objectives for plant transformation which are thought to be biologically possible and economically desirable. Where conventional genetic methods can be used to modify plants, they will often be easier and so preferable to methods based on gene isolation and transfer. However, the latter may prove to be the only means for transferring genes from widely different organisms. Table 7.1 shows which of the different techniques of plant molecular biology can presently be applied to each of the three crops covered by this report.

Plants which have been genetically engineered by recombinant DNA technology to introduce a desired gene or trait will need to be subjected to rigorous testing in the same way as the segregating progenies from conventional breeding programmes. Further, the exploitation of particular new or altered genes may require adjustment of the general genetic background which can only be achieved by using the genes in conventional breeding programmes.

For crop plants in general, four broad areas can be recognised as particularly amenable to study by molecular biological methods. These are:

Photosynthesis. Among plants having the C3 photosynthetic metabolism there is much variation in photosynthetic characteristics. It may be presumed that individual wild species have become optimally adapted to their environment as a result of evolution over thousands, if not millions of years. Species grown as crops, though intensively selected by man, may not be optimally adapted in terms of their photosynthetic characteristics, for selection has been mainly for other characters. Although knowledge of the biochemistry and biophysics of photosynthesis is con-

Table 7.1. *Status of the development of the techniques of molecular biology*

Technique (already established for one or more plant species)	Refer to Section	Application to		
		Wheat	Oilseed rape	Faba beans
Preparation of genomic libraries, cloning, gene isolation	3.1	Yes	Yes	Yes
Sequencing of genes	3.1	Yes	Yes	Yes
Detection of genes in genomic libraries	3.2			
(a) given the protein product of the gene		Yes	Yes	Yes
(b) given the mRNA		Yes	Yes	Yes
(c) by transposable elements, given a dominant allele of a gene		No	No	No
Location of gene to chromosome and position on chromosome, with the aid of 'classical' genetic mapping procedures	3.2			
(a) given the phenotypic expression of the gene		Yes	Yes	Yes
(b) given the protein product of the gene		Yes	Yes	Yes
(c) given the mRNA		Yes	Yes	Yes
Construction of new genes *in vitro* with required activity (i.e. appropriate control sequences)	3.3	Yes	Yes	Yes
Insertion of genes into plant cells and regeneration of plants from these cells	3.4	No	Reported	No
Deletion or inactivation of existing genes	5.5.1	No	No	No
Location of foreign genetic material in host tissue (e.g. viral pathogen, rye segments in wheat cultivars)	4.6.2	Yes	Yes	Yes

siderable it is not possible with confidence to identify limiting components. *Molecular biology, by allowing particular changes to be made, makes it possible for the first time to answer questions about limitations imposed by component processes. Results from research to answer these questions will indicate what scope there is for improving crop production by modifying the processes of light harvesting and photosynthetic carbon metabolism.*

Plant water relations. Mesophytes require means of restricting their water loss whilst maximising carbon dioxide uptake, and a variety of mechanisms for achieving this has evolved, depending on the droughtiness of the environment. It is of world-wide importance to explore means of improving the drought tolerance of crop plants and molecular biology can help to achieve this by identifying genes responsible for the adaptive responses of plants to water stress as well as those whose expression is continuous (i.e. does not depend on water stress). *Studies on selected wild species might reveal genes which, if transferred to crop plants, would improve their drought tolerance without unacceptable deleterious effects. The most obvious process for study is osmoregulation, which allows the maintenance of turgor and hence photosynthesis and growth at reduced water potentials.*

Seed composition. There is much genetic variation within crop species in the gross composition and in the composition of different fractions. Within wide limits, this genetic variation does not affect fitness. Therefore, it can be exploited, with no penalty in terms of reduced yield, pest or disease resistance, to satisfy man's requirements. *For the relatively 'new' crop, oilseed rape, there is need for considerable further change, in terms of reduced linolenic acid, glucosinolate and phospholipid content. For faba beans, not at present a major crop in the EEC, molecular biology could be used to improve the methionine content of the protein they contain.*

Pest and disease resistance. Conventional breeding has for many crops enabled varieties to be produced which have a satisfactory level of resistance to pests and diseases. Because of genetic change in the pests and pathogens, maintaining the resistance requires continued effort by breeders. Yet, most plant species are effectively resistant to most pests and pathogens. *It is tempting, therefore, to suppose that resistance genes could be transferred from alien species to crop species to give much more durable resistance than hitherto available.* The chief difficulty with this approach is the problem of identifying genes, or their products, which render the

Table 7.2 *Needs for the application of molecular biological methods to the improvement of wheat*

Character/gene	Use/benefit*	Comment	Refer to Section
Related to yield			
Photoperiod genes	Rapid identification in progenies of particular alleles giving desired life cycle duration and developmental pattern	Major genes known	4.1
Vernalisation genes		Major genes known	4.6.1
Precocity genes		Genes believed to exist but not satisfactorily identified	
Genes affecting photosynthetic rate**	Because of low heritability of photo-synthetic rates, identification of desired alleles in progenies will be much more efficient than selection for gas exchange rates or other measures of photosynthesis	No genes yet identified which show allelic variation. Metabolic control points not not yet established	4.6.1
Genes affecting water use efficiency	Improved water economy and drought resistance	No genes yet identified	4.1 4.6.1
Major genes affecting plant height	Ability to identify known genes without the need to make test crosses. Improved yield and lodging resistance	Several major genes known, two loci of major importance for yield	4.1 4.6.1
Disease resistance			
Resistance to powdery mildew and yellow rust	(a) Recognition of race-specific resistance genes in host cultivars	Technology available	4.2.1 4.2.2 4.6.2

		Technology available	
Resistance to all economically important diseases of wheat	(b) Recognition of virulence genes in pathogen (possibly within the host tissue) Identification of durable resistance and the genes coding for it	Basic research needed. If successful, would be of great value in breeding	4.6.1 – 4.6.5 4.6.2
Pest resistance Resistance to shoot flies		Search for resistance mechanisms which could be transferred from other species to wheat, transfer could be made when a transformation system exists for wheat	
Grain quality Endosperm protein variants important for quality**	More rapid identification of allelic variants	Genes known, allelic variation described for the two major storage proteins	4.4 4.6.4

** Only the genes for these characters have been isolated.
* When insertion of new genes into wheat is possible, directed modification of the characters can be undertaken, by changing the genes *in vitro*, before insertion.

host plant totally resistant to a 'foreign' pest or pathogen. *In many cases, the resistance will probably be due to non-recognition or to the presence of a 'positive' defence system. Thus attention should be focussed on identifying recognition systems and those conferring defence.* Both are likely to be active at surfaces, and so could be constituents of the cuticle, cell wall, or (particularly for pests) volatile substances.

In addition to the general points mentioned above, the following have been identified as objectives for the crops covered in this report which can best be tackled by the methods of molecular biology:

Wheat (Table 7.2). Until a vector and suitable transformation system has been developed, the transfer of alien genes to wheat by molecular methods cannot be made. Work to overcome these problems is urgently required. In the meantime, rapid methods of identifying genes conferring major gene resistance to rusts and mildew should be developed. Also, the molecular basis of good quality in the glutenin and gliadin proteins requires further study, so that geneticists can select for transfer to bread wheat only those proteins which are the most likely to confer improved quality. Further work is needed to establish the molecular basis of the high photosynthetic rates of some wild diploid species so that specific genes for transfer to bread wheat can be identified.

Oilseed rape (Table 7.3). A system based on cDNA clones should be developed for identifying S-incompatibility alleles in seedlings. This would greatly aid the production of lines with appropriate S alleles, which could be used in the production of F_1 hybrid rape.

The possibility of transferring to oilseed rape genes from the related genus *Moricandia*, with the objective of reducing photorespiration and hence increasing net photosynthesis, should be investigated.

The molecular basis of resistance to stem canker (*Leptosphaeria maculans*) needs to be established so that the genes concerned can be isolated from *B. juncea* (a resistant species) and transferred to *B. napus*.

General resistance to the numerous insect pests of oilseed rape (many are common to other Brassicas) should be sought and transferred to oilseed rape.

Chemical or radiation- induced mutagenesis should be tried as a means for reducing the glucosinolate content of the seed, whilst retaining levels in the vegetative tissue which may be necessary to maintain resistance to insect pests.

Table 7.3 *Needs for the application of molecular biology to the improvement of oilseed rape*

Character/gene	Use/benefit*	Comment	Refer to Section
Related to yield			
Resistance to safe, broad-spectrum herbicide(s)**	Improved, simpler and cheaper weed control	Existing atrazine resistance is of limited value. Work could be rapidly outdated by development of improved herbicides	5.1 5.5.1
Self-incompatibility alleles, as part of breeding and seed production**	Rapid detection in seedlings of S-alleles would aid the exploitation of self-incompatibility in breeding	Few alleles available in *B. napus*. More could be transferred from other *Brassica* species	5.5.1
Disease resistance			
Resistance to major diseases	Means for identifying disease resistance genes so that they can be handled more effectively in breeding programmes	Basic research needed	5.2 5.2.2
Pest resistance			
Resistance to large group of 'Brassica specialist' pests	Decreased cost of pest control	Basic research needed. Resistance probably needs to be introduced from other, unrelated genera	5.3 5.5.3
Seed quality			
Steps in fatty acid biosynthesis	Improved oil quality		5.4 5.5.4

** Only the genes for these characters have been isolated.

* Insertion of genes into oilseed rape is now possible, and directed modification of characters can now be undertaken, by changing the genes in vitro, before insertion.

Faba beans (Table 7.4). The gene pool available to breeders of this crop is very limited and ways of introducing variation from related species into faba beans are urgently required. From the point of view of yield, improved stability of yield over sites and seasons and over years, as well as improved yield potential, are required. This might be achieved by improving the drought resistance of the crop by the introduction of alien variation. More stable cytoplasmic male sterility (CMS) would enable an effective system for producing F_1 hybrids to be developed. A molecular analysis of CMS in '447' cytoplasm should provide a basis for achieving this.

Improved resistance to the fungal pathogens *Ascochyta* and *Botrytis* is required and may have to be introduced from other *Vicia* species by gene isolation and transfer using a vector.

The value of faba beans for animal feeding is limited by their relatively low protein content in comparison with oilseed cakes (especially soya cake) and by deficiency in methionine. Work is needed to assess the scope for increasing the protein content by increasing the number of copies of the storage protein genes. To improve the methionine content, it may prove possible to modify the amino acids in the variable regions of the legumin and vicilin storage proteins, by substituting methionine for histidine residues, and the feasibility of achieving this should be investigated.

Table 7.4 *Needs for the application of molecular biology to the improvement of faba beans*

Character/gene*	Use/benefit**	Comment	Refer to Section
Related to yield			
Increased winter hardiness	Ability to grow the winter form of the crop in 'continental' European climates	Would have to be introduced from other genera	6.1 6.5.1
Improved autofertility	Would permit pedigree breeding systems and make crop less dependent on pollinating insects	Genes not identified	6.5.1
Improved cytoplasmic male sterility	Would make the production of F_1 hybrids a practical possibility	Present source of male sterility is inadequate	6.5.1
Methods for introduction of alien genetic variation from other *Vicia* species and related genera	Potentially large benefits in terms of yield, quality and pest and disease resistance	Available gene pool presently limited to *V. faba*	6.1 6.5.1
Disease resistance			
Resistance to major diseases	Identification of genes for resistance and their more rapid exploitation in breeding	Basic research needed	6.2 6.5.2
Pest resistance			
Resistance to *Aphis fabae*	Decreased use of pesticides	Good resistance exists in other *Vicia* species	6.3.1 6.5.3

Table 7.4 (cont.)

Character/gene*	Use/benefit**	Comment	Refer to Section
Seed quality			
Removal of antinutritional factors (tannin, L-DOPA, vicine, convicine)	Improved feeding value	Variation might be generated by mutagenesis	6.4 6.5.4
Increased lysine and methionine content of protein	Improved feeding value	Basic studies required of structure and variability of legumin and vicilin	6.4 6.5.4

* None of the genes have been isolated.
** When insertion of new genes into faba beans is possible, directed modification of the characters can be undertaken, by changing the genes *in vitro*, before insertion.

REFERENCES

de Block, M., Schell, J. and Van Montagu, M. (1985). Chloroplast transformation by *Agrobacterium tumefaciens*. *EMBO Journal,* **4**, 1367–72.

Fincham, J.R.S. and Sastry, G.R.K. (1974). Controlling elements in maize. *Annual Reviews of Genetics*, **8**, 15–50.

Pelletier, G., Primard, C., Vedel, F., Chetrit, P., Remy, R., Rousselle, P. and Renard, M. (1982). Intergeneric cytoplasmic hybridisation in Cruciferae by protoplast fusion. *Molecular and General Genetics*, **191**, 244–50.

FURTHER READING

Chapters 2 and 3 Crop improvement by breeding and Molecular biology and plant breeding

Albersheim, P. and Anderson-Prouty, A. J. (1975). Carbohydrates, proteins, cell surfaces and the biochemistry of pathogenesis. *Annual Review of Plant Physiology*, **26**, 31–52.

Bushnell, W. R. and Roelfs, A. P. (1984). *The cereal rusts, Vol. 1. Origins, specificity, structure and physiology*. Orlando: Academic Press.

Callow, J. A. (1977). Recognition, resistance and the role of plant lectins in host parasite interactions. *Advances in Botanical Research*, **4**, 1–49.

Comai, L. and Stalker, D. M. (1984). Impact of genetic engineering on crop protection. *Crop Protection*, **3**, 399–408.

Flavell, R. B. (1985). DNA transposition – a major contribution to plant chromosome structure. *BioEssays*, **1**, 21–2.

Gleba, Y. Y. and Sytnik, K. M. (1984). *Protoplast fusion*. Berlin: Springer-Verlag.

Harborne, J. B. (1977). *Introduction to ecological biochemistry*. London: Academic Press.

Heath, M. C. (1981). A generalised concept of host–parasite specificity. *Phytopathology*, **71**, 1121–3.

Johnson, R. (1984). A critical analysis of durable resistance. *Annual Review of Phytopathology*, **22**, 309–30.

Keen, N. T. (1982). Specific recognition in gene-for-gene host–parasite systems. *Advances in Plant Pathology*, **1**, 35–82. London: Academic Press.

Paleg, L. G. and Aspinall, D. (1981). *The physiology and biochemistry of drought resistance in plants*. Sydney: Academic Press.

Rosenthal, G. A. and Janzen, D. H. (1979). *Herbivores: their interaction with secondary plant metabolities*, New York: Academic Press.

Rudulier, D. Le, Strom, A. R., Dandekar, A. M., Smith, L. T. and Valentine, R. C. (1984). Molecular biology of osmoregulation. *Science*, **224**, 1064–8.

Slonimski, P., Borst, P. and Attards, G. (1982). *Mitochondrial genes*. New York: Cold Spring Harbor Laboratory.

Vanderplank, J. E. (1978). *Genetic and molecular basis of plant pathogenesis*. Berlin: Springer-Verlag.

Chapter 4 Wheat

Austin, R. B. (1982). Crop characteristics and the potential yield of wheat. *Journal of Agricultural Science, Cambridge*, **48**, 447–53.

Austin, R. B. and Jones, H. G. (1975). The physiology of wheat. In Report of the Plant Breeding Institute for 1974, pp. 20–73.

Austin, R. B., Morgan, C. L., Ford, M. A. and Bhagwat, S. G. (1982). Flag leaf photosynthesis of *Triticum aestivum* and related diploid and tetraploid species. *Annals of Botany*, **49**, 177–89.

Austin, R. B., Ford, M. A., Morgan, C. L., Kaminski, A. and Miller, T. E. (1984). Genetic constraints on photosynthesis and yield in wheat. *Advances in Photosynthesis Research*, Vol. IV, ed. C. Sybesma, pp. 103–10.

Bennett, F. G. A. (1984). Resistance to powdery mildew in wheat: a review of its use in agriculture and breeding programmes. *Plant Pathology*, **33**, 279–300.

Flavell, R. B., Payne, P. I., Thompson, R. D. and Law, C. N. (1984). Strategies for the improvement of wheat-grain quality using molecular genetics. *Biotechnology and Genetic Engineering Reviews*, **2**, 157–73.

Gale, M. D. and Youssefian, S. (1985). Dwarfing genes in wheat. In *Progress in plant breeding*, ed. G. E. Russell, pp. 1–35. London: Butterworth.

Innes, P., Blackwell, R. D. and Quarrie, S. A. (1984). Some effects of genetic variation in drought-induced abscisic acid accumulation on the yield and water use of spring wheat. *Journal of Agricultural Science, Cambridge*, **102**, 341–51.

Lambers, H. (1982). Cyanide-resistant respiration: a non-phosphorylating electon transport pathway acting as an energy overflow. *Physiologia Plantarum*, **55**, 478-85.

Law, C. N., Worland, A. J. and Giorgi, B. (1976). The genetic control of ear emergence time by chromosomes 5A and 5D of wheat. *Heredity*, **36**, 49–58.

Payne, P. I., Holt, L. M., Jackson, E. A. and Law, C. N. (1984). Wheat storage proteins: their genetics and their potential for manipulation by plant breeding. *Philosophical Transactions of the Royal Society, London, B*, **304**, 359–71.

Pugsley, A. T. (1971). A genetic analysis of the spring–winter habit of growth in wheat. *Australian Journal of Agricultural Research*, **22**, 21–31.

Scarth, R. and Law, C. N. (1984). The control of daylength response in wheat by the group 2 chromosomes. *Zeitschrift für Pflanzenzuchtung*, **92**, 140–50.

Staniforth, A. R. (1979). *Cereal straw*. Oxford: Clarendon Press.

Chapter 5 Oilseed rape

Anon. (1983). Proceedings of the Sixth International Rapeseed Conference, Paris, Vols. 1 and 2, 1782 pp.

Anon. (1984). Agronomy, physiology, plant breeding and crop protection of oilseed rape. *Aspects of Applied Biology*, **6**, London: Association of Applied Biologists.

Bunting, E. S. (Ed.) (1981). *Production and utilisation of protein in oilseed crops*. The Hague: Nijhoff.

Buzza, G. (1983). The inheritance of an apetalous character in Canola (*Brassica napus*). *Eucarpia Cruciferae Newsletter*, **8**, 11–12.

Fenwick, G. R., Heaney, R. K. and Mullin, W. J. (1983). Glucosinolates and their breakdown products in food and food plants. *CRC Critical Reviews in Food Science and Nutrition*, **18**, 123–201.

Kramer, J. K. G., Sauer, F. D. and Pigden, W. J. (Eds.) (1983). *High and low erucic acid rapeseed oils*. Toronto: Academic Press.

Nasrallah, M. E., Dowey, R. C. and Nasrallah, J. B., (1983). Biochemical genetic analysis of self incompatibility in *Brassica*. In *Pollen biology and implications for plant breeding*, ed. D. L. Mulcahy and E. Ottaviano, pp. 251–6. New York: Elsevier.

Scarisbrick, D. H. and Daniels, R. (Eds.) (1986) *Oilseed rape*. London: Collins.

Scott Holaday, A., Tyrone-Harrison, A. and Chollet, R. (1982). Photosynthetic/photo-respiratory CO_2 exhange and characteristic of the C_3–C_4 intermediate species, *Moricandia arvensis*. *Plant Science Letters*, **27**, 181–9.

Thompson, K. F. (1983). Breeding winter oilseed rape, *Brassica napus*. *Advances in Applied Biology*, **7**, 1–86.

Weiss, E. A. (1983). *Oilseed crops*. London: Longman.

Chapter 6 Faba beans

Bond, D. A. (Ed.) (1980). *Vicia faba: feeding value, processing and viruses. Proceedings of an EEC seminar on research on the improvement of plant proteins.* The Hague: Nijhoff.

Dantuma, G., Von Kittlitz, E., Frauen, M. and Bond, D. A. (1983). Yield, yield stability and measurements of morphological and phenological characters of faba bean varieties grown in a wide range of environments in Western Europe. *Zeitschrift für Pflanzenzuchtung*, **90**, 88–105.

Rexen, F. and Munck, L. (1984). *Cereal crops for industrial use in Europe.* Report prepared for the Commission of the European Communities (EUR 9617 EN).

Hebblethwaite, D. (1983). *Vicia faba research in Europe. Report prepared for the Commission of the European Communities* (EUR 8649 EN).

Hebblethwaite, P. D. (Ed.) (1983). *The faba bean – a basis for improvement.* London: Butterworth.

Lawes, D. A. (1973). The development of self fertile field beans. Report of the Welsh Plant Breeding Station for 1972, pp. 163–76.

Matta, N., Gatehouse, J. A. and Boulter, D. (1981). The structure of legumin of *Vicia faba* L. – a reappraisal. *Journal of Experimental Botany*, **32**, 183–97.

Nielsen, N. C. (1984). The chemistry of legume storage proteins. *Philosophical Transactions of the Royal Society, London, B,* **304**, 287–96.

Stoddard, F. L. (1986). Pollination and fertilization in commercial crops of field beans (*V. faba* L.). *Journal of Agricultural Science, Cambridge*, (in press).

Witty, J. F., Minchin, F. R. and Sheehy, J. E. (1983). Carbon costs of nitrogenase activity in legume root nodules determined using acetylene and oxygen. *Journal of Experimental Botany*, **34**, 951–63.